THE CITY

HALF A CENTURY OF MUNICIPAL DECLINE, 1935–1985

POLITICS OF THE CITY

HALF A CENTURY OF MUNICIPAL DECLINE, 1935–1985

MARTIN LOUGHLIN, M. DAVID GELFAND &
KEN YOUNG

Routledge
Taylor & Francis Group
LONDON AND NEW YORK

First published in 1985

This edition published in 2007
Routledge
2 Park Square, Milton Park, Abingdon, Oxon, OX14 4RN

Simultaneously published in the USA and Canada by Routledge

711 Third Avenue, New York, NY 10017

Routledge is an imprint of Taylor & Francis Group, an informa business

Transferred to Digital Printing 2007

First issued in paperback 2013

The publishers have made every effort to contact authors and copyright holders of the works reprinted in the *The City* series. This has not been possible in every case, however, and we would welcome correspondence from those individuals or organisations we have been unable to trace.

These reprints are taken from original copies of each book. In many cases the condition of these originals is not perfect. The publisher has gone to great lengths to ensure the quality of these reprints, but wishes to point out that certain characteristics of the original copies will, of necessity, be apparent in reprints thereof.

British Library Cataloguing in Publication Data
A CIP catalogue record for this book
is available from the British Library

ISBN13: 978-0-415-41750-1 (volume hbk)
ISBN13: 978-0-415-41930-7 (subset)
ISBN13: 978-0-415-41318-3 (set)
ISBN13: 978-0-415-86035-2 (volume pbk)

Routledge Library Editions: The City

First published in 1985

This edition published in 2007
Routledge
2 Park Square, Milton Park, Abingdon, Oxon, OX14 4RN

Simultaneously published in the USA and Canada by Routledge

711 Third Avenue, New York, NY 10017

Routledge is an imprint of Taylor & Francis Group, an informa business

Transferred to Digital Printing 2007

First issued in paperback 2013

British Library Cataloguing in Publication Data
A CIP catalogue record for this book
is available from the British Library

ISBN13: 978-0-415-41750-1 (volume hbk)
ISBN13: 978-0-415-41930-7 (subset)
ISBN13: 978-0-415-41318-3 (set)
ISBN13: 978-0-415-86035-2 (volume pbk)

Routledge Library Editions: The City

HALF A CENTURY OF MUNICIPAL DECLINE, 1935–1985

MARTIN LOUGHLIN, M. DAVID GELFAND &
KEN YOUNG

Routledge
Taylor & Francis Group
LONDON AND NEW YORK

Half a Century of Municipal Decline 1935–1985

edited by

Martin Loughlin
London School of Economics and Political Science

M. David Gelfand
Tulane Law Schoo

Ken Young
Policy Studies Institute

with a Foreword by J. A. G. Griffith

London
GEORGE ALLEN & UNWIN
Boston Sydney

George Allen & Unwin (Publishers) Ltd,
40 Museum Street, London WC1A 1LU, UK

George Allen & Unwin (Publishers) Ltd,
Park Lane, Hemel Hempstead, Herts HP2 4TE, UK

Allen & Unwin, Inc.,
Fifty Cross Street, Winchester, Mass. 01890, USA

George Allen & Unwin Australia Pty Ltd,
8 Napier Street, North Sydney, NSW 2060, Australia

First published in 1985

British Library Cataloguing in Publication Data

Half a century of municipal decline, 1935–1985.
1. Local government – Great Britain – History – 20th century
I. Loughlin, Martin II. Gelfand, M. David
III. Young, Ken, *1943–*
352.041 JS3091
ISBN 0-04-352118-5
ISBN 0-04-352120-7 Pbk

Library of Congress Cataloging in Publication Data

Half a century of municipal decline, 1935–1985.
Bibliography: p.
Includes index.
1. Local government – Great Britain – History – 20th century – Addresses, essays, lectures. 2. Municipal government – Great Britain – History – 20th century – Addresses, essays, lectures. I. Loughlin, Martin. II. Gelfand, M. David. III. Young, Ken.
JS3091.H34 1985 352.041 84–28405
ISBN 0-04-352118-5 (alk. paper)
ISBN 0-04-352120-7 (pbk.: alk. paper)

Set in 11 on 13 point Imprint by Computape (Pickering) Ltd
and printed in Great Britain by
Mackays of Chatham

Contents

Foreword

In the earliest days, local government was its own justification. Even as the central sovereign power grew, local government remained local self-government and the distinction between local government and local administration had little meaning outside the judicial process of the king's courts. In local matters the judicial and the administrative functions were often in the same hands.

The development of the royal power brought its conflicts with local government which were settled and transformed by the outcome of the constitutional crises of the seventeenth century. It is not too great a simplification to say that those conflicts were not a serious political issue for the next 150 years. Our modern world began in the 1830s. My copy of the *Everyman Encyclopaedia* under the title 'Local Government' expresses some concern about fundamentals:

> Perhaps the most remarkable feature of modern local government is the rapid advance in municipal trading. England is in advance of most other countries in this respect, and it is becoming axiomatic to say that no one yet knows either how far such policy will be carried in the future and to what extent it may be said to be in accordance with sound canons of political economy. It is so directly antithetical to the doctrine of *laissez-faire*, and such a complete subversion of the time-worn notion that state functions end with the conservation of internal order and the defence against aggression from without, that many stigmatise it with the name of municipal socialism and look with apprehension on the rapid increase of rates and municipal debt . . . . The increase within the last thirty years is unquestionably due partly to the progress in municipal trading enterprise, a feature of public life which is regarded by Lord Avebury, and no doubt with strict historical accuracy, as entirely opposed to the liberal teaching of Cobden, Mill and Bright . . . . Experience, however, shows that municipalities are not embarrassed by the added duties involved in providing and managing tramways, railways, omnibuses, steamboats, and fire insurance, the supply of electrical fittings, clothing, confectionery in school, milk and eggs, and in brickmaking and tailoring

> .... Judged by the light of the evolution of modern Liberalism the whole question is really not so much a matter of profit and loss, or the effect of municipal or private enterprise, as one of opinion whether or no the interests of civilisation are better served by the state becoming its own shopkeeper and factory owner instead of allowing the private individual or company to exploit other individuals or companies for gain. Thus stated, it becomes clear that the issue is between every conceivable form of individualism, absolute or modified, and that co-operative mode of adjusting human relations which for want of a better name is called Socialism.

Those words were written in 1913, almost exactly halfway between 1835 and the present. They reflect the great new impetus given to local authorities by the reforms of the late nineteenth century and the impact of the ideas of the early Fabians. Looking back from 1985, reformers like Jennings, Laski and Robson seem part of that progressive movement rather than of later radical, even revolutionary, tendencies; they saw a thriving local government as part of the democratic defence against fascism. Yet local government failed to expand beyond its traditional functions and so began its decline. I suppose we may say that housing between the wars and from 1945 to 1955, and educational administration after 1945, were major advances in municipal development, and that in neither of these was central policy control unduly onerous. Even in the early 1930s the Ministry of Health rejected the drastic advice given by the Ray Committee on Local Expenditure. The loss of functions under the Attlee government, however justifiable on other grounds, was serious because it restricted the range of local government activities. But there was plenty to do after 1945. To the reformers, the new London was a great advance and even the 1972 changes produced those large metropolitan counties which looked as though they might enable local government to become regional authorities for planning (economic as well as land-use) in the places where this was most required.

Local government was still not in serious danger of being reduced in status to local administration but it certainly was no longer its own justification, and so was laid open to attack. When governments seek to draw more and more power into their own hands, sooner rather than later they come into conflict with autonomous bodies. The

Environment Secretary, Patrick Jenkin, has recently made much play with the idea of Britain as a unitary state, but he is wrong. It is not only in federations that power may be plurally distributed. The powers of municipalities derive historically from bargains struck between them and the royal central power, with advantages to both sides. Such bodies are likely to prove difficult when their autonomy is threatened. James II's failure to suborn the municipalities contributed greatly to his own destruction when with a folly amounting to genius he managed in three years to throw away his kingship. Commenting on this episode a hundred years ago, the great historian Ranke wrote:

> One of the principal internal reasons for the existence of corporate rights might be sought for precisely in the doctrine that the whole life of the state ought not to depend on the tendencies of the sovereign power which are fluctuating by their very nature; there must be somewhere a limit to their momentary influences beyond which a free development of municipalities and, under their protection, of individuals remains possible.

And twenty years earlier, another historian, Hallam, spoke of this interference as the most capital delinquency of the king's government because it directly attacked the fundamental constitution of the state.

All this is no less true today than it was three hundred years ago. At the heart of the constitution of the United Kingdom is a tension, a tautness, that ultimately determines all the institutional relationships within the state. It can be expressed in many ways: as an aspect of the sovereignty of Parliament, a reflection of party politics, an incident in the relationship between front-benchers and back-benchers, a pressure on the judicial process, a facet of representative democracy, a fundamental part of the notion of legality itself. It touches on individual freedom, on ministerial accountability, on the redress of grievances, on voting supply, on the limits of authority, on the stability of the state itself.

Certainly, the present government has its majority and so the political capacity to make new laws depriving the municipalities of their independence. But the absence of a written constitution does not set the government free to take any action it pleases to deprive municipalities of their historical powers. If it does so, it destroys that tension between the central executive and the local electorate. No doubt the state could be run without elected councillors. Local officers could be transformed into administrative agents of the central government. Schools would continue to exist, planning regulations to be enforced, houses to be built, fires to be put out. But the character of our parliamentary democracy would be altogether altered.

It seems to me that the provisions for rate-capping in the Rates Act 1984 are little removed from a proposal to replace elected councils by administrative units. For a very long time, local inhabitants have enjoyed the right to elect local representatives with the power to tax and so to determine, within modest political limits, what level of services shall be provided in the locality. Hundreds of Acts of Parliament have conferred this right directly on the local electorate, and these powers on locally elected councils. Local government has been a part of the constitutional structure of government and of politics, for rather longer perhaps than there has been a recognizable central government.

Moreover, the history of the powers of local authorities has not been the history of central government conferring powers through the medium of Parliament. Until the middle of the last century, these powers were commonly obtained through local legislation in local Acts with which governments frequently had little or nothing to do.

To say of a statute that it is legally unconstitutional is meaningless. All statutes are legally constitutional. But I have no difficulty in saying of an Act to put a limit on the rates leviable by a local authority that it is politically unconstitutional. By that I mean it runs counter to the spirit and letter of the existing constitution, and makes the change quite fundamental. This is not merely an incident in an attempt to control public expenditure. It deprives people in the localities of a real exercise of power, and transfers it to the central government.

The contributions to this book trace the changes of the last fifty years and it is impossible to disagree with their general verdict of decline. Does this matter? Should we regard the transformation of local government into local administration as serious? This is a separate question from that concerning the reduction of the quality of local services due to shortage of finance, although that too must always be serious. If it could be shown that locally elected councill-ors, in a substantial proportion of cases, came to decisions which, in terms of the quality of services provided, were worse than those taken by ministers, the argument for the retention and strengthening of local government would be weak. There is, however, little evidence to support such a proposition. We can certainly say that many of the best developments since 1935, as before, have been initiated by local authorities. And we can certainly say that many of the worst – such as industrialized building – have been forced on local authorities by central government departments. At the least, the record is ambiguous. The overriding consideration remains: that local government involves the activities of thousands of people, locally elected and locally appointed. That still provides the ultimate democratic justification for the preservation and enlargement of local government.

What, then, of the next fifty years? Will a comparable volume of essays recording the situation two hundred years after 1835 be entitled *The Decline and Fall of Local Government*? I think not. The present crisis has invigorated local government in the worst affected areas as campaigns and activities particularly in Greater London, Liverpool, Sheffield and the metropolitan counties have shown. It is also a commonplace to say that the United Kingdom is becoming more and more divided between the relatively rich south and the relatively poor Midlands and north. This will, in its turn, make for a greater regionalization and is likely to promote stronger local loyalties. Local politicians in those poorer areas are becoming more outspoken and more powerful; and they see their responsibilities as centred on their localities and against the central government in London. So economic and political forces are strengthening those local authorities even at this time when they are under the greatest pressure. And out of the present discontents, local government will, I believe, emerge as a powerful institution and once again become its own justification.

London School of Economics
and Political Science JOHN GRIFFITH

Preface

In 1935 George Allen & Unwin published *A Century of Municipal Progress, 1835–1935*, edited by Harold Laski, Ivor Jennings and William Robson. The book was published under the auspices of the National Association of Local Government Officers (NALGO), and its appearance was timed to coincide with the celebration of the centenary of the Municipal Corporations Act. The tone of *Municipal Progress* (as we shall call it) was optimistic and self-confident. It recorded achievements across the entire range of local government services and anticipated achievements yet to come. 'It is appropriate', remarked the editors, 'that on the occasion of the centenary of the first Municipal Corporations Act we should estimate the changes which a century of development in local government have brought about' (p. 12). It seems equally appropriate for us to take stock of the changes which have occurred since.

This present volume, written half a century later, records a different story. Further 'progress' there has been in the sense of the enhancement of welfare arising from developments in public service provision. But in some instances the stream of improvement has been diverted and expansion reversed. It is not our purpose, however, to estimate the consequences for welfare of these changes in service provision, unforeseen as they were in 1935. Rather, we have concentrated on the governmental process, on the role and scope of local government and its place in the national framework of public policy. In choosing this narrower focus we can perhaps justify our adoption, as a title for this book, of a simple inversion of the original.

In many fundamental respects it has indeed been a half-century of decline; a decline so accelerated by the actions of the present government as to raise vital questions about the nature of the polity itself. In other respects it is over-simplistic to speak of 'decline', just as it was, fifty years ago, to speak of 'progress'. In terms of service provision it has been a half-century not of decline but of transformation, of social and economic change so comprehensive as to preclude any simple set of comparisons between local government in 1935 and 1985. Perhaps the most striking transformation is the growth in the size and significance of local government as an employer. Despite the loss of a number of major services to national bodies in the postwar period, the number of local government employees in Great Britain has almost trebled over this fifty-year period and during the postwar period alone local authority employment as a proportion of the total working population doubled.

A related change of significance is the progressive unionization of this workforce, which reached a figure of 85·6 per cent in 1974. In 1935 when *Municipal Progress* was written, NALGO had 80,000 members. It had around 350,000 in 1965; by 1976 the membership had reached 683,000. Between 1965 and 1976 the membership of NUPE (the main manual worker union) rose from 248,000 to 650,500. The two major local government unions (not all of whose members work in local government) are now among the largest affiliated to the Trades Union Congress. The significance of these developments for industrial relations in local government lies beyond the scope of this book. But one feature which does deserve attention is the transformation of NALGO's role during the fifty-year period, symbolized by the 1964 decision to affiliate to the TUC after forty-two years of hesitation and debate, and the holding of the first strike by any group of NALGO members that same year.

These developments, and the greater assertiveness of the public sector unions generally, would have surprised the editors of the 1935 volume. For them, NALGO was a public service body, which 'recognises not merely its duty to its own members but also its duty to the public .... The fact that such a public spirit prevails among local government officers is one of the main reasons for our belief in the great future of English local government' (pp. 11–12). But the NALGO of which they wrote, and which sponsored the production of *Municipal Progress*, has since moved far from its origins. NALGO's early history was dominated by chief officers and by such nationally influential Conservatives as Sir Herbert Blair and Sir Homewood Crawford. Throughout much of its history it was characterized, in Robson's own words, by 'a sense of public service, of *noblesse oblige* towards the community'. But the adoption of a strike clause in the NALGO constitution in 1964 drove a wedge between the chief officers as guardians of such values and the larger membership. The chief officers claimed that their professional status and ethics and their special responsibility to their employers and to the community made striking impossible for them. They were therefore given exemption from the requirements of solidarity and NALGO then became an integral part of the wider trade union movement.

NALGO's history and development provides an appropriate starting-point for this book. The origins of *Municipal Progress* are to be found in NALGO's campaign to protect local services from expenditure cuts in the early 1930s; the origins of this present volume lie in our concern to understand the nature of the current attempts by central government to constrain and control local government fifty years later. The two books originate in similar crises: the desire of governments to subordinate the institutions of local democracy to a particular and contentious vision of the national interest.

Our collective purpose in writing this book has been primarily to explain rather than to defend, and in particular to explain how the long-run trends and structural weaknesses, illuminated as they are by a deeper historical perspective than is usual, have led to the present confrontation. We first began to explore these issues in a small discussion group convened by David Gelfand and meeting in Oxford. Our intention was to enable public lawyers and political scientists to develop a common approach to issues of current importance. Martin Loughlin first called attention to the significance of the 'jubilee' of *Municipal Progress* and from then on, in a series of larger meetings organized by Ken Young at the Policy Studies Institute and the London School of Economics, we drew in economists and social policy analysts to work with us in the preparation of this book.

From the beginning we recognized that we could not aim to replicate the breadth of coverage of the original volume. Nor would it be useful or appropriate to do so. We have focused instead on the larger picture – the influence of social, economic, financial, legal and political developments of the last half-century on the institution of local government and the more significant shifts in the scope and functions of local government itself.

Of these shifts the most notable have occurred in the fields of land-use planning, housing and social welfare and we therefore chose to devote separate chapters to each, at the admitted expense of other vital services: these three examples best illustrate the nature and impact of the broader trends of the last half-century. Consequently, because the basic settlement of 1944 has remained intact (although under considerable strain today) we have not devoted a separate chapter to education, although, since it is the most important service in expenditure and manpower terms, education issues are dealt with as they arise in the general chapters of the book.

Finally, we should like to record our gratitude to NALGO for supporting the publication of this volume and thereby bringing it within the reach of a wider audience. We also wish to acknowledge the financial support provided by the Economic and Social Research Council to enable a workshop to be held in July 1984 at which preliminary versions of these papers were presented. The production of the book has been a collective effort, involving the contributors in a series of meetings and providing for mutual support and criticism. A number of other people attended some of these meetings, joined in the discussion of the papers and gave us the benefit of their advice. Of these, we should particularly like to record our gratitude to George Jones, who has also read and commented upon parts of the manuscript, and to Susan Barrett, Chris Game, John Hills, Jim Sharpe and Stella Theodoulou. We are also grateful to Mary Alden for assistance with the organization of meetings and the preparation of this manuscript, and to Susan Johnson for help with the bibliography.

Coventry, New Orleans, London — M.L.
September 1984 — M.D.G.
K.Y.

1

Re-Reading the Municipal Progress: a Crisis Revisited

KEN YOUNG

To every text a context: the purpose of this chapter is to establish the context within which *Municipal Progress* was written. This exercise might seem at first glance unnecessary; the book purported to be, and self-evidently was, a celebration of the centenary of the Municipal Corporations Act 1835. Yet there remains the question of *why* these authors, in their surprising association with NALGO, felt the centenary to mark so significant a point in the history of local government in England. The answer to that question takes us far beyond the centenary itself into broader areas of concern, and into the political and intellectual milieu of the mid-1930s.

To understand this wider setting is to see *Municipal Progress* as a document for its time, an ephemeral text to which few will turn today for an authoritative account of the development of local government. If it deserves re-reading fifty years later it does so less on grounds of its merits as a social and administrative history but for the parallels between that time and our own. *Municipal Progress* is at root a defence of local government as an institution and an attempt to restate a basic premise: *that local self-government is indispensable to the democratic polity*.

Writing in 1935, the editors, if not all the contributors, were concerned to make three fundamental points. First was the need to defend local government as a provider of often-overlooked public services against the panic over public expenditure which followed the 1931 financial and political crisis. Second was the view that the steadily increasing scope of local government in an urbanizing society reflected the development of civilization itself, a process characterized by the growth and improvement of cities and by the gradual assumption of public responsibility for matters hitherto thought to lie in the private domain. Third was the belief in an organic connection between the existence of a vigorous system of local self-government and the maintenance of free democratic institutions at the national level.

To re-read *Municipal Progress* is to reconsider the large issues: the good life, the city beautiful, the maintenance of political freedom. They are introduced to the reader on the opening page, where the editors quote with approval Graham Wallas's dictum that 'for the average English citizen the possibility of health, of happiness, of progress to the old Greek ideal of "beautiful goodness" depends on his local government more than on any other factor in his environment'.

Other contributors to this present volume examine the extent to which the assumptions and predictions of 1935 have been rebutted during the half-century that followed. That Laski, Jennings, Robson and their collaborators were so wrong about the future in so many particular respects should surprise no reasonable person; nor should it lead us to dismiss them. On the larger questions they were perhaps more nearly right and for this reason *Municipal Progress* can be re-read today with sobering and salutary effect.

This chapter looks first at the attempts to achieve local expenditure cuts during the period 1931–3 and shows how the opposition to the economy drive grew into a broader promotional campaign, one product of which was *Municipal Progress* itself. It then deals with publication and reception of the book before exploring the wider intellectual and political concerns which the authors shared.

The Immediate Context: the Attack on Local Government

Written as an ostensible celebration of the Municipal Corporations Act 1835, *Municipal Progress* was, as a project, rooted in the economic and political crisis of 1931 and the economy campaign which followed it. For the first time in modern English history the institutions of local government were threatened with a fundamental subordination to national policies on public expenditure and a sharp reversal of at least fifty years of their extension.

The Great Depression of 1929 had brought about a catastrophic fall in demand for British exports and triggered substantial increases in unemployment to more than 2.5 million by the end of 1930 (Ashworth, 1960; Skidelsky, 1967). The recently elected minority Labour government stumbled from crisis to crisis and, bereft of other than conventional ideas and helpless in the face of rising unemployment, fell victim to its own financial orthodoxy. As Chancellor of the Exchequer Philip Snowden faced an increasing deficit with expenditure on unemployment relief mounting and tax revenues falling. Snowden 'believed fanatically that a balanced budget was the greatest contribution which a government could make to overcoming the Depression, and did not hesitate at the resistance this would provoke in his own party' (Taylor, 1965, p. 286). He appointed a committee to examine and recommend economies in national expenditure, chaired by Sir George May of the Prudential Assurance Company.

Snowden imagined that the report of the May Committee would cow both his own supporters into accepting cuts in public expenditure and the Conservative Party into accepting increased taxation. Published on 31 July 1931, its principal effect was rather to split the Cabinet, with its advocacy of expenditure cuts of £96 million, to be obtained by 20 per cent reductions in unemployment relief and in public sector pay. A run on the pound in early August brought forward consideration of the report, as the Bank of England urgently sought dramatic gestures from the government in the interests of restoring international confidence.

It soon became apparent that the government could not survive. The TUC General Council declared its opposition to cuts, and almost half the Cabinet held out against the reductions in unemployment relief. On 24 August 1931 Ramsay MacDonald formed a National Government 'to deal with the national emergency', his ten-man Cabinet containing four Conservative, four Labour and two Liberal ministers (Bassett, 1963). The way was now clear for the acceptance in early September of Snowden's emergency budget, which imposed cuts of £70 million, and for MacDonald to seek a 'doctor's mandate' in the general election held in late October, which resulted in the temporary annihilation of the Labour Party and a new phase of Conservative dominance of the House of Commons.

Apart from the reductions in unemployment relief, the expenditure cuts of September 1931 fell heavily upon public sector salaries and wages (Treasury, 1931). A general cut of 10 per cent was to be made in the remuneration of all of those paid by the state, from Cabinet ministers downwards (Taylor, 1965; Gilbert, 1980). The May Committee had proposed a 20 per cent reduction in teachers' pay and 12.5 per cent in that of the police (May, 1931). The new government moderated these cuts in the first instance; then on 15 September came the Invergordon 'mutiny' when men of the fleet refused duties in protest at the cuts in the pay of naval ratings, and further 'revisions' were made.

Where the cuts should fall and with what weight was a matter of interdepartmental bargaining, but the Cabinet minutes for these troubled weeks also document a recognition that inequitable cuts might prove to be unenforceable, even before Invergordon confirmed that fear. At the forefront of the concern to achieve equity lay the question of local government pay. Local authority staff were not servants of the state and, with the exception of teachers and the police whose pay came in part from central government grant, lay outside the remit of the May Committee and beyond the direct influence of the government itself. Could cuts in police and teachers' pay be realistically proposed if other local government staff were uniquely exempt from the scheme?

The Cabinet thought not. At an early pre-budget Cabinet meeting the Home Secretary Sir Herbert Samuel reported on his meetings with the Police Council where the point had been made that:

> The police, like the fighting services, could only be induced to accept the cuts in pay and pensions provided it was necessary for the salvation of the country, and provided all branches of the public service made equivalent sacrifices. (CAB 23/68: 31 August 1931)

The police suggested that the Minister of Health should instead consider cuts in local government pay. A week later the teachers made the same point with equal force; the Cabinet recognized that 'an anomaly would arise' (in the delicate prose of the Cabinet secretary) 'if the teachers sustained cuts in their salaries while a highly paid Director of Education suffered no cut' (CAB 23/68: 7 September 1931). A similar point was made as to the 'anomalies' arising from chief constables being the only local authority chief officers to suffer reductions.

But how were such cuts to be made? The local authorities – or at any rate some and in all probability most of them – could be induced to propose reductions, but staff were effectively protected against their enforcement by the terms of their contracts of employment. Neville Chamberlain, Minister of Health, accordingly proposed to the Cabinet that the government might take the power 'to terminate the contracts between local authorities and their staffs'; in this way the authorities 'might be empowered by legislation to cancel contracts so to enable them to negotiate new ones'. This extraordinary idea 'was discussed at considerable length both in its political and practical aspects'; which is to say it was decisively shelved. Local authorities were to be allowed 'to deal with the question as best they could' and to seek reductions by agreement. 'Anomalies', however painful and potentially dangerous, would have to be tolerated (CAB 23/68: 7 September 1931).

The government's immediate response to the May Report was to send a circular to the local authorities, exhorting economies. But here the government faced the problem not only of being largely reliant upon its persuasive powers, but also of the lack of uniformity in pay practice between authorities, which impeded any general formula for standard reductions:

> The conditions of the local government service and the ranges of salary vary so materially that HMG do not think it practicable, even if it were desirable, for them to impose any hard and fast rule on local authorities in this matter. But they are confident that officers of the local government service not affected by the reductions in emoluments of nationally financed services will be prepared to make their contribution in the present emergency, and they think that each local authority should discuss the situation with its officers with the object of ensuring that all have an opportunity of sharing equitably in the sacrifices demanded by national need. (Ministry of Health, 1931)

With this circular began the chain of events that led to NALGO's campaign to defend local government staff against compulsory 'sacrifices' and, eventually, to promote the cause of local autonomy and local service provision. For the moment NALGO viewed the minister's advice without undue alarm, although William Robson was incensed by the circular: the choice of the term 'impose' in relation to the salary reductions was, he complained, 'a strange word for the central government to employ towards the ancient counties and proud cities of England' (Robson, 1933, p. 88).

In the event, as Levi Hill recalled in *Municipal Progress* (pp. 139–40), only about 80 per cent of officers became subject to deductions, and these on a temporary one-year basis; some local authorities chose to ignore the minister's exhortations. The lower-paid officers were virtually exempt from cuts while the highest paid suffered no more than a cut of 10 per cent. Most of the pay reductions ran for twelve months from early 1932 but some were agreed rather later; the last few schemes were due to expire in 1935.

Perhaps the most remarkable feature of this first post-crisis attempt to reduce local authority expenditure was the lack of vigour with which the government pursued it. Whereas the May Committee saw a need to achieve permanent reductions in public expenditure the politicians and officials saw the crisis simply as that: a short-lived financial exigency (May, 1931, p. 222). The Ministry of Health circular, so mild in tone, had been readily agreed by the Cabinet. Not only had the proposal to abrogate contracts of employment been decisively rejected, but also the Cabinet was sensitive to the need 'to safeguard the position of local government officials, whether against excessive cuts in their salaries, or even dismissal' (CAB 23/68: 7 September 1931). Indeed, in contrast with the situation a half-century later, the central government in 1931, for all its apparent concern for economy, effectively stood between the local government staff and their employers.

The Cabinet sought a balanced budget to overcome a short-term financial crisis, and showed little concern to bring about a general and lasting reduction in public expenditure. Locally, however, a number of Conservative-led authorities sought to achieve a once-and-for-all reduction in manning and services. The political balance shifted in their favour after the October general election when Chamberlain became Chancellor of the Exchequer and the initiative in local expenditure questions passed with him from the Ministry of Health to the Treasury.

Chamberlain, in common with subsequent chancellors, encountered the gulf between central government's desires and the possibility of their achievement through the action of local authorities. Not only was the centre ill-equipped to manipulate local expenditure other than through grant reductions and the curtailment of loan sanction, but also it was hardly in a position to know local circumstances; the Chancellor could bridge that gap only by securing the collaboration of local government itself, through the local authority associations.

On 1 July 1932 Chamberlain wrote to the local authority associations to inform them that 'in considering what further action can be taken to lighten the heavy burden at present borne by the community HMG have given attention to the question of a reduction in the expenditure of local authorities'. While recognizing that many authorities had taken steps to reduce expenditure, the government still felt that 'further action, alike early and decisive, is requisite' and with that aim in view invited the associations to establish a committee 'to consider the whole field of local expenditure'; it was a matter of 'grave importance and urgency' and the government would give all assistance to the committee by servicing it, providing information and underwriting the costs of the inquiry (Ray, 1932).

Thus was conceived the first determined assault on local expenditure. The new Committee on Local Expenditure appointed as its chairman Sir William Ray, Conservative leader of the LCC and newly elected MP for Richmond, and under his guidance the majority of the committee set about the task of reversing the postwar growth of expenditure. Their position was very clear: they had witnessed 'a growth in local government expenditure and local government commitments which must call for very serious thought and indicate that a very heavy burden has been placed upon the private citizen' (para. 10). Its reduction would not only lift that burden and 'set money free for the development of trade and industry but, by removing anxieties lest even greater burdens should be imposed, will also stimulate business enterprise and encourage private individuals in wise expenditure' (para. 5). These were more ambitious goals than those of the Ministry of Health although doubtless they were shared by the Treasury. Moreover, they might have some resonance in the country at large. The Anti-Waste campaign that followed the Geddes cuts in 1921 had shown itself a potent force electorally, and one as hostile to Conservative as to Socialist interests (Young, 1975, pp. 129–37). A revival of that campaign was not unimaginable.

Local expenditure had not grown by linear progression since the war. A total rate levy of around £113 million in 1919–20 had risen steeply in the following year to around £150 million and stayed at about that level through the decade. Although capital expenditure had quadrupled from the base of £24 million in 1919–20 by ten years later, it had stayed on a plateau throughout most of the 1920s at around the 1929–30 level. On the other hand central government grants in aid of revenue expenditure *had* grown steadily, from £48 million in 1919–20 to more than twice that sum ten years later; and debt burdens had grown in like proportion. Although these last were not readily adjustable, they suggested that the growth in local expenditure had been encouraged and permitted by central government.

Accordingly, while making many specific proposals in their 173-page report on the opportunities for local economies, the Ray Committee mounted a fundamentalist attack on the covert conspiracy between central and local officials to encourage growth and expansion. Like the May Committee, Ray sought not mere emergency measures, but 'permanent measures of economy' (para. 4) and to that end called upon central government to 'refrain from placing fresh burdens on the country' (para. 10); they declared that 'the time has come when a halt should be called to legislation involving local authorities in fresh expenditure' (para. 11); they advocated that working-class housing 'cease to be a charge on the public purse' (para. 99); they attacked the system of percentage grants in aid of local expenditure as encouraging 'too great an extension of services' (para. 11) and called instead for a block grant system (para. 13). They also expressed 'considerable anxiety' over the administration of the Poor Law and argued that local government faced 'disaster' through the political manipulation of out-door relief; their conclusion was that it should be removed from local government altogether (paras 223–6). These were matters of high policy, and constituted a major challenge to the government. Moreover, their arguments were often heard outside the major parties and could not be divorced from the wider political challenge that was to be mounted after 1931 by the 'third force', the growing independent and ratepayer movement (Young, 1975, pp. 151–69).

In so overreaching themselves Sir William Ray and his colleagues in fact ensured the defeat of the economy movement which they sought to lead. They stimulated the energetic counter-attack led by NALGO and supported by Laski, Jennings, Robson and their friends. In claiming so much the Ray Committee won very little. It is notable that, while published as a Command paper, the Ray Report was not the subject of a House of Commons debate; that the Cabinet put to one side the major question of scope of state responsibility so impertinently raised by the committee; and that while the committee had been requested to report by October 1932 (and had done so) not until March 1933 was the government's response considered by the Cabinet, by which time the report had been overtaken by new turns in government policy.

Although the apparent uncontrollability of local government pay had been a central concern of the emergency measures of 1931, the Ray Committee had little new to suggest twelve months later. They found no general statistics on the costs of local government salaries and wages (para. 232) and were hampered by the absence of any national system (para. 236). Reviewing the impact of the 1931 circular, the committee had found 'no complete information as to what has been done and it seems that there has been a substantial variation in practice'. They considered, however, that 'so far as the wide differences in bases of payment admit, all authorities should endeavour to make reductions on a comparable basis' (para. 237). Yet they failed to make any positive suggestions for achieving such consistency, despite making specific proposals for economies estimated to save £5 million on salaries during 1932–3, and up to £40 million on all expenditure after three or four years.

From Resisting Economies to Celebrating Progress

As the major organizing body for local authority white-collar staff, NALGO, with its membership of around 80,000 officers, had not been much perturbed by the 1931 crisis. Local government officers were cushioned against government action and they regarded with equanimity Chamberlain's circular. The cuts were temporary, implemented by agreement and avoided where possible. The stance taken by the Ministry of Health was seen as little more than a gesture. The threat posed by economy-minded authorities was far greater, however, and in accepting the Treasury invitation to appoint members to the Committee on Local Expenditure the local authority associations had given implicit support to the fundamentalist attack on the role of central and local government in national life.

The Conservative-dominated counties were solidly behind the Ray Report, resolving that:

> it is the duty of all County Councils, without foregoing the right to criticise detailed proposals when brought forward by the Departments concerned, to aid in giving effect to the fullest practicable extent to the policy of the expenditure committee. (NALGO, 1933a)

There is no doubt that some authorities used the Ray Report as an excuse to postpone commitments, particularly when encouraged to do so by confusingly presented government guidelines on loan sanction (Ministry of Health, 1933). As ever, William Robson proves a colourful, if perhaps unreliable, guide to the impact of the Ray Report on local decision-making:

> with shirt-fronts bursting with patriotic fervour, councillors and aldermen filled their fountain pens and prepared to slaughter the local estimates. Thousands of items, large and small, ranging from great schemes such as the Humber bridge and the Dartford–Purfleet tunnel, sanctioned by Parliament and costing millions of pounds, to little routine replacements of worn-out equipment, were suspended or cancelled. (Robson, 1933, p. 87)

NALGO did not wait for such events to unfold. When the Committee on Local Expenditure was announced in July 1932 the NALGO executive immediately appointed a three-man emergency subcommittee to ensure that the association was fully consulted and to take such steps as might be necessary to influence public opinion against 'excessive economy'. The subcommittee later recalled that 'terrible barriers had to be broken down in the first case, and an almost universally unsympathetic press to be persuaded' (NALGO, 1933a). NALGO embarked upon a series of initiatives designed to reverse the unfavourable public perception of local government. They instituted a speaking campaign using prominent MPs (including, ironically, Philip Snowden, who was no longer in the government) and launched a major letter-writing campaign aimed at achieving favourable press coverage. The emergency subcommittee gave evidence to the Ray Committee, putting forward a general argument as to the wider benefits of social expenditure, a point which the committee chose to regard as beyond its terms of reference (Ray, 1932, para. 5). But despite the considerable effort and not inconsiderable expenditure involved in NALGO's initial 'fight back' campaign, the membership were unimpressed by the results and there was some criticism of the value of the public relations consultant who had been recruited for the campaign (Spoor, 1967).

NALGO's general secretary, Levi Hill, consistently sought a wider approach to the campaign, which he saw as concerned to promote the image of local government in general, rather than to resist specific immediate economies. As one who was closely involved in NALGO's educational activities and in the efforts to raise professional standards in public administration he was closer than many of his associates to those academic experts who were sympathetic to local government. He now sought to enlist their aid. Robson was already associated with NALGO and wrote occasional pamphlets – *The Arithmetic of the Rates* (1932) was one – for popular circulation. Laski was still more closely involved, having served on the Hadow Committee on local government officers and having been co-opted to the emergency subcommittee at Hill's suggestion as early as July 1932. For Laski, the campaign was a 'great experiment' to disperse 'clouds of misinformation' about local government and, by fostering more positive attitudes to its development and extension, 'open a new epoch in the history of British democracy' (Spoor, 1967, p. 134).

In March 1933 Hill reported the results of his attempts to involve a wider range of 'University professors and other persons interested in local government' in the publicity campaign. Among these figures were Sir Ernest Barker (Laski's Oxford tutor) and G. D. H. Cole, both of whom were sympathetic in a general fashion but could offer nothing specific; J. M. Keynes who was 'overwhelmed with work'; John Hammond who was 'over-occupied and over-taxed [*sic*]'; John Maud, who was ill and unable to help; and a number of others. Positive offers came from Professor T. Simey of Liverpool (a former local government officer), who suggested the publication of articles evaluating the benefits of the social expenditure of local government, and from Ivor Jennings who also offered the help of the *Local Government Chronicle*. Jennings put forward an extensive list of worthwhile initiatives, including articles on such topics as 'how rates are spent' (which he subsequently wrote and republished in booklet form), a series of articles covering the period 1833–1933 and a series of lectures to be held at the LSE (NALGO, 1933b). Laski had already pointed out the significance of the centenary of the Municipal Corporations Act due in 1935 and preliminary arrangements to celebrate it had already been made when Jennings's proposal for a series of commemorative articles was tabled.

Whereas NALGO's stance in 1932 was one of vigorous defence of the local government system, the abrupt change in the political climate early in the following year enabled the association to take the initiative and recapture the high ground of the 'economy' debate. That they were able to do so owed much to the position of the Ministry of Health. While the Ray Committee was castigating the growth of local government expenditure the ministry was setting it in a longer-term context (Ministry of Health, 1933). Whereas the trend towards greater indebtedness fed the committee's anxieties, the sanguine ministry extolled the loan debts of local authorities as 'comparable with . . . the capital of a commercial undertaking' and for the most part relatively insignificant in fiscal terms (Ministry of Health, 1932, p. 150). For the ministry it was a matter of satisfaction that 1932 could be judged one of 'progress in a number of services vitally affecting the life and well-being of the people'.

Sir Hilton Young, who had replaced Chamberlain as Minister of Health, defended this position in Parliament and implicitly rejected Ray's arguments by justifying public expenditure in other than strict book-keeping terms:

> there are other ledgers than the cash ledger – the ledger of health, the ledger of amenities, the ledger of education. Values cannot always be estimated in cash and the social services must be reckoned in the same way by their uses to the community . . . the development of the assets of the nation are not measurable in money. (Quoted in Ministry of Health, 1933, p. 157)

With these words the Ray Report was smothered before its dangerous doctrines could take too firm a hold on Conservative thinking.

NALGO officials seemed aware at an early stage in their campaign that the Minister of Health and his officials, having already stood between the local authorities and their staff, also shared NALGO's views and values on public expenditure. The reception by ministry officials of a NALGO deputation on the Ray proposals was certainly warm, with reference being made by Sir Arthur Robinson, the Permanent Secretary, to the views expressed in NALGO's evidence to Ray, views which the committee had not chosen to consider (NALGO, 1933c).

The Ray Committee having reported in October 1932, NALGO sought to ensure that the contrary standpoint was heard before the government determined its course of action. The considerable delay before the report came to Cabinet helped defuse its proposals; in January 1933 NALGO minutes record that 'the general attitude towards the demand for a reduction in expenditure on local services had undergone a considerable change' and that 'drastic recommendations to local authorities were no longer expected' (NALGO, 1933a).

Such was indeed the case. The immediate budgetary crisis over, the government had little interest in rolling back the frontiers of public responsibility in the manner envisaged by Sir William Ray. The ministry view – contrary to that of the Treasury – was that local authorities had complied 'readily' with the exigencies of a temporary situation and were 'alert to the need for a policy of *wise economy* and for reducing to the *lowest figure consistent with the discharge of local government functions*, the burden of local rates' (Ministry of Health, 1932, p. 160; my emphases).

Sir Hilton Young took his proposed response to the Ray Report to Cabinet on 8 March 1933, his officials being engaged that same day in reassuring NALGO with the prediction that the circular 'would do no more than direct the attention of local authorities to the recommendations contained in the Committee's report' (NALGO, 1933c). Young meanwhile was telling the Cabinet that the report contained 'various suggestions for better and more careful administration which he proposed to commend to the local authorities'. The Cabinet doubted the wisdom of referring to Ray's proposal that the possibilities of a block grant system should be subject to further independent inquiry; there was little desire to respond to Ray's proposals on Poor Law administration, for it touched the nerve of conflict between economy and 'political considerations'; and Young thought it hardly necessary even to mention in the circular that the specific recommendations of the committee 'were superseded by [his own] recent pronouncements' (CAB 23/75: 8 March 1933).

The Ray Report, dead within months of its inception, was formally buried in July 1933. Sir Hilton Young, while applauding the fall in the average rate collected per head from £3 15s in 1930–1 to £3 12s in 1932–3 and paying tribute to the Ray Committee (which could in no way have influenced this particular reduction), pointed out that 'economy is not an abstract matter in itself, but is one aspect of efficiency'. He claimed that very full account was taken of the Ray Report 'and the vast majority of its recommendations were at once commended ... to the local authorities' (House of Commons, 1933b). The minister's suavity would have done little to mollify the campaigning member for Richmond. Only the previous day Sir William Ray had attacked once again the principle of percentage grants and claimed 'that we have not carried economy far enough':

> It is a dangerous doctrine to preach in these days but two years ago we were all enthusiastic about economy. Some of us tried to carry it out and I very much regret that it has not been carried out to a much larger extent by His Majesty's Government and by many of the local authorities in this country.

His committee had been representative and largely unanimous, he claimed; and 'if the government wanted some strength behind it in regard to economy, there it was'. 'I have finished my little rebuke', concluded the disappointed economizer, 'I have been saving it up for some time' (House of Commons, 1933a). Sir William Ray's fundamentalism had something of the shrillness of the apostate: as leader of the Conservatives at London's County Hall he had steered the LCC into an expansionist housing programme in 1930 on the grounds that without such social expenditures the party would be electorally vulnerable (Young and Garside, 1982, p. 163).

In June 1933 NALGO President Sam Lord (a borough treasurer by profession) told his conference that they had survived 'two years of intense depression in public administration unprecedented in the history of the country'. However, he continued,

> the efforts of those who sought to imperil the existence of our local government institutions have suffered a severe set-back, and in my opinion these efforts have helped consolidate the resolve of the general public that any undue interference with services vital to the well-being of the community will not be tolerated. The misguided so-called economy campaign has failed. (NALGO, 1933d)

For NALGO the emergency was over, and in July the essentially defensive emergency subcommittee was dissolved into a more positive public relations committee to press home the association's fortuitously won advantage.

The pamphleteering continued, with Robson, Jennings, Simey and Kent Wright producing pamphlets for NALGO. The centenary, meanwhile, 'provided a heaven-sent focus' for the new campaign (Spoor, 1967, p. 136). NALGO headquarters planned and encouraged the widest celebration, rightly claiming that 'but for NALGO, that anniversary would have passed unnoticed'. There was a national essay competition on the topic 'what I owe to local government'; a banquet at the Guildhall; 60 local authorities arranged a total of 174 local events and 26 issued centenary booklets; 800 articles on the centenary in newspapers and journals were recorded (NALGO, 1935). The King, whose silver jubilee fell that same year, was sent a telegram extolling the virtues of local government and the happy coincidence of the two anniversaries.

Organizing the celebrations put a heavy strain on the association, remarked NALGO President G. W. Coster, but

> the indispensable work of local government and some idea of its value as a national institution controlling the welfare of this country was brought to a point which has never before been reached . . . we have made a very substantial contribution to the uplift of local government. (NALGO, 1936)

The fetes and banquets were ephemeral. Of all the events planned by NALGO to promote Levi Hill's idea of local government as a vital institution 'with the local government officer its pillar' none was as significant as the preparation and publication of *A Century of Municipal Progress*. The various proposals by Laski and Jennings (and doubtless Robson too) in 1932 coalesced in the idea of a major volume of essays, taking the LSE lectures which Jennings had initiated as a starting-point.

Municipal Progress and its Authors

Municipal Progress is by today's standards a substantial volume, running to 511 pages, and including an extensive chronological table of key developments in local government history, compiled by C. Kent Wright, the Town Clerk of Stoke Newington Metropolitan Borough Council. The book was published by Allen & Unwin at the fairly high price of one guinea, whereas Robson's polemic volume

The Government and Misgovernment of London (1938) was subsequently to be published by the same house at around a third of that price. Much of what *Municipal Progress* contained had been rehearsed in the public lectures 'to celebrate the local government centenary' given at the London School of Economics during the period January to March 1935. The three editors took part in the lecture series, as did J. L. Hammond, Herman Finer, Lord Snell and Sir Robert Ensor, only the last of whom did not subsequently contribute to *Municipal Progress*.

The editors assembled a team of distinguished contributors. The three editors were themselves among the giants of their disciplines at that time. Ivor (later Sir Ivor) Jennings was Reader in Law at the LSE, a leading scholar in the field of public law and political science and editor of the *Local Government Chronicle*; he is best remembered today as the author of what remained for many years standard works on the law and practice of the constitution. Harold Laski had succeeded Graham Wallas as Professor of Political Science at the LSE and did much to maintain the radical reputation of that institution; he combined a prodigious output of work across the entire field of intellectual history and political science with an active political involvement which brought him chairmanship of the Labour Party in the 'high tide' year, 1945. William Robson, reportedly Laski's first student at LSE, was then editor of the influential *Political Quarterly* and Reader in Administrative Law at LSE (where he was to continue to teach public administration until shortly before his death in 1980); another prolific writer in the fields of political science and administration, he was the most fiercely polemical of the three.

The names of several of their contributors have a resonance today: Herman Finer, another LSE political scientist who was to achieve an international reputation in subsequent years at Harvard; John Hammond, one of the leading social historians of his age, joint author with his wife Barbara of the series of classic studies of the labourer in English history; Sir George Newman, the formidable former Chief Medical Officer of the Ministry of Health; Sir Ernest Simon (later Lord Simon of Wythenshaw), former Liberal MP and junior minister and the driving force in Manchester's interwar housing programme; Elie Halevy, Professor at the Ecole Libre des Sciences Politiques in Paris, remembered for his work on philosophical radicalism; Sir Josiah Stamp (later Lord Stamp of Shortlands), the public finance economist, statistician and member of numerous public and private boards; Herbert Morrison (later Lord Morrison of Lambeth), the leading Labour politician who had only recently shown Labour the road back to power with the party's striking victory at the 1934 London County Council elections, a past Minister of Transport, a future Foreign Secretary and Deputy Prime Minister.

The remaining contributors are less well known to today's reader. They included Lord (Henry) Snell, the former farm labourer who progressed via Nottingham University and the LSE to become chairman of the LCC, MP for East Woolwich, a junior minister at the India Office and to hold prominent positions in international and imperial organizations. This energetic figure (who recorded his recreations as 'none' in *Who's Who*) eventually became deputy leader of the House of Lords in Churchill's wartime coalition. Few outside NALGO will today remember Levi Hill, General Secretary of NALGO from 1909 to 1943 although, as we have seen, Hill played a vital role in the promotion of local government in the interwar years. Other contributors included Frank Smith, Professor of Education at Leeds University; Stanley Jast, the City of Manchester's chief librarian; Sir Frederick Kenyon, Director of the British Museum; Gilbert Slater, the former principal of Ruskin College; and John Willis, a young law scholar who had recently published his doctoral thesis on the supersession of private Acts by public general legislation as the principal form of parliamentary regulation of local government.

Little point would be served in reviewing the broad territory they traversed. Subsequent chapters in this volume make sufficient reference to the particular arguments and viewpoints expressed by these seventeen authors in the pages of *Municipal Progress*. The essential character of the book is to be found in the editorial contributions; Laski, Jennings, and above all Robson (whose concluding essay casts a somewhat reckless glance forward to 2035) between them define the central issue: the role of local government in the democratic polity. In this task they were supported more readily by Snell than by the other contributors. Snell, in his essay on 'The Town Council', extends the span of the argument both temporally (to 'the spiritual soil of Athens, Rome and Jerusalem') and spatially (to the imperial expression of British civilization). Laski develops Snell's rather breathless examination of local self-government with a closely focused study of the working of the committee system in local government; Robson and Jennings both deal with central–local government relations and Robson's dissection of the weaknesses of the existing structure of local government is paralleled by Willis's portrayal of the growing legislative role of the Whitehall departments.

Municipal Progress does not seem to have been widely reviewed. Notwithstanding the stature of the editors it was received less as a major contribution to learning than as a milestone, recording how far we had travelled since 1835, and how much further we might need and expect to go. If generally it was politely received, the critics found little that was new; the historical background was largely drawn from the Webbs; Robson based much of his own concluding chapter on his *Development of Local Government* (1931a). Laski's examination of the committee system was picked out both by contemporary reviewers (Beales, 1936) and, subsequently, by Robson himself (Robson, 1950) as the most illuminating in the book. Here Laski exploited his unique qualities as both a political scientist of unrivalled breadth of vision and as an active local politician ('Comrade Alderman Professor Laski' as fellow Fulham Council member Michael Stewart, 1980, p. 40, remembers him).

Leonard Beales's full-length review article, appearing in Robson's *Political Quarterly*, was somewhat ambivalent and self-confessedly 'ungracious'. Beales's major criticism is worth recalling:

> We can applaud achievements, admirably recorded in this book, but how much surer we should be about their origin and their maintenance, and how much more effective in securing their enlargement, if we knew more about such matters as the occupations of councillors, the processes and motives of their selection both as candidates and as members of local authorities, about the comparative effectiveness of different authorities, and other related questions of social politics, or political sociology, or communal psychology or whatever – short, one trusts, of social pathology – this subject of research should be called. It is unfair, perhaps, to ask for the results of largely unmade inquiries in that dim borderland where political power is generated. Yet without disciplined inquiry in that field, there is a danger that superficial conclusions will masquerade as whole truth. (Beales, 1936, p. 101)

Such a deficiency strikes today's reader. But another thirty years were to pass before the neglect of the issues which Beales identified began to be rectified by a new generation of scholars. It should be recognized that *Municipal Progress* was less a work of political science than one of political argument.

Read as political argument, one further feature of the book attracts particular attention: the suppression, in what is essentially a celebratory volume, of the deep reservations about the then current capacity of the local government system to live up to the demands which the editors foresaw the next century would make upon it. True, Robson's closing chapter rehearses the criticisms of the present arrangement of area and power already displayed in his *Development of Local Government* (1931a) but the tone is admonitory rather than hectoring. Nevertheless, Robson regarded his *Development of Local Government* as 'a devastating criticism of the English municipal system as it exists at present' and had earlier been left 'speechless' by the 'outrageous unfairness' of a *Spectator* reviewer describing him as 'a fervent admirer of the present system of local government' (Robson, 1931b).

There is no outrage, no devastation to be found in the pages of *Municipal Progress*. This was a celebration, a panegyric to 'the great tradition of adaptation and reform'. The withering criticisms which Robson was usually all too ready to level at local government (in contrast to Jennings, the sympathetic editor of the *Local Government Chronicle*) would have spoiled the mood of the party. It would have been out of place in a volume that sought not just to celebrate but to defend and promote local government in the face of the then-current threats to its scope, standing and autonomy.

That these fears were reasonable when the book was conceived – they had receded somewhat by the time it appeared – is something which today's reader will recognize for, as Alexander's chapter in this volume shows, the last decade has produced a threat as profound as that which *Municipal Progress* was designed to resist. *Municipal Progress* was, like NALGO's other public relations initiatives, born of necessity in the resistance to a profound if short-lived threat to the status, autonomy and integrity of the local public service. By the time it appeared, three years after the idea was first mooted, the nature of the threat had changed. There were larger issues to face and more critical battles to fight.

The Intellectual in Politics: Civilization under Threat

In the key chapters of *Municipal Progress* there is a noticeable preoccupation with the nature of civilization. 'The city councils' services', proclaimed Sir Herbert Simon, 'mean the difference between savagery and civilisation' (p. 11). Lord Snell elaborated: 'the influence of English local government upon the life of the nation . . . has been profound' and has produced a reservoir of administrative capacity 'on which British civilisation may continue to draw' (p. 81). It would be wrong to dismiss these references as rhetorical flourishes. Within *Municipal Progress* is to be found a clear understanding of the nature of that civilization. It rested upon three foundations: a special link of continuity with the classical tradition; an organic connection with urbanism and urbanity; and a gradual drawing into the public realm of matters hitherto considered to be private concerns.

As to the first, several of the eminent authors of 1935 were the young men of the late Victorian era. They shared that peculiarly unendearing Victorian quality, an unquestioned adoption of classical antiquity as a reference point for the present; in particular of 'ideas and institutions which were not really derived from Greece but to which some Victorians chose to give a Hellenic colouring' (Jenkyns, 1980, p. ix). Accordingly, classical references abound. The English towns in 1835, remarks Hammond, 'would have put a citizen of the Roman empire to the blush. They had none of the amenities, few of the decencies of civilisation' (p. 37).

Snell seemed incapable of making a point without classical reference, but the substance of what he sees in the local government centenary is barely distinguishable from Robson's own vision and perhaps from Laski's too. Snell's somewhat narcissistic essay is only the most explicit statement of a shared conception of a modern renaissance of the older virtues:

> Speaking of a section of the citizens of Athens, Thucydides said that 'they were born into the world to take no rest themselves and to give none to others' and something of the same kind might be said of many of those who in England in our own day devote their lives to its local government service. The average town councillor may, unlike the citizen of Athens, have no conscious theory of the State, and no knowledge of the civic idealism of ancient Greece;

> but he, like them, can fall in love with the city in which he dwells and be sustained by an ideal the nature of which he doesn't fully apprehend. He lives in the present and has a passion for efficiency; but the roots of his moral and social life are deeply set in the spiritual soil of Athens, Rome and Jerusalem. (p. 81).

Perhaps the ex-farm labourer Snell was led to overemphasize the unique role of the Great City in the maintenance of virtue. But he was not alone, for the vision of the authors of *Municipal Progress* is an urban vision to a degree that even the centenary itself – a centenary of *urban* reform – can hardly justify. There was something of a consensus among urbane Europeans and their intellectual associates; as a comparable study of local government in the British colonies puts it:

> The identification between local government, or (in more general terms) government at short range within small communities and the growth of civilisation has been too uniform and too frequently repeated not to be significant, 'Cities and culture and civilisation', writes Mr. Jawaharlal Nehru, 'seem to go together . . . with the growth of cities learning also grows, and the spirit of freedom'. (Hinden, 1948, pp. 4–5)

Compare the different American tradition with its Thoreauesque suspicion of the city, so well drawn together by Morton and Lucia White (1962).

Municipal Progress emphasized the movement from a concern with public health and civic works in the nineteenth century to the broader concern with culture, learning, the arts – with *the city beautiful* – in the twentieth. In more than a literal sense, the Victorian City (not itself much appreciated by Robson) laid the infrastructure for the Great Cities that were to be celebrated in his later writings. The social, rather than the physical, infrastructure was the more significant. Living at closer quarters, city-dwellers generated needs and mutual obligations to which the extension of law and of the public domain were inevitable responses. Because spill-over effects are endemic in residential contiguity, the regulation of 'nuisance' and the demand for public goods provision are inherent in the urban setting. Urban change was seen in 1935 as a process of growth and agglomeration; the scope of government in regulating and providing could be conceived of only as progressively increasing. On this view civilization is a process, not an end-state. It takes the form of a linear extension of *the realm of the public*. Accordingly, Laski, Jennings and Robson looked forward to the further inexorable growth of local government as the embodiment of both components of the civilizing process.

Robson was the most explicit about this process as it was manifested through the growth of government, while Laski was more concerned with the regulation of the non-governmental. For Robson the growth of government, irrespective of the regime, entailed the expansion of the realm of law and the greater ordering of human behaviour, 'subordinating private interests to the public good'. Robson explored this particular significance of law as a pattern of conduct in his *Civilisation and the Growth of Law* (1935c), published in the same year as *Municipal Progress*.

Laski's preoccupations were broader: for him, socialism was less about law than about the growing self-regulation of pluralistic institutions in service of the public good. In *Reflections on the Revolution of our Time* (1943) he envisaged a 'planned democracy' in which formerly defensive associations – the Bar Council, the British Medical Association – would be liberated from the burdens of self-interest. Reassured by 'the economics of abundance' they 'will find that their protective function becomes far less important than the contribution they can make to the improvement of standards of legal and medical practice' (p. 351) – a process of enlightened self-denial that Laski foresaw extending as far as the National Union of Mineworkers. In this romantic vision we find the key to Laski's warm support for NALGO: surely here was the prototypical case of the self-regulation of the collective interest of a class of workers in the public good? Certainly the editors of *Municipal Progress* thought so as they commended NALGO's example to their readers:

> a professional organisation, it rightly asserts, must concern itself with the advancement of the public interest in its particular sphere of action. (p. 12)

NALGO's demonstration of the possibility of a 'corporate public spirit' sustained Laski and Robson's faith in local government and in a socialist pluralism.

In much of this, the contributors of *Municipal Progress* echoed that body of thinking with which the name of Norbert Elias is most directly associated. Elias drew the closest connections between the growth of the state in the later Middle Ages and the development of notions of civilized behaviour: the development of a sense of greater human interdependency, expressed in changing customs and in the taking into public control of private and intimate acts. The social regulation of behaviour at table and in the bedroom, the development of courtesies, conventions about the bodily functions and the control of aggression all facilitated the development of a social *order* upon which the modern state came to be founded through the 'dynamics of human interdependence' and the eventual monopolization of force. Elias published *The Civilising Process* in Basle as *Über den Prozess der Zivilisation* in 1938; he finished it in 1936 after a period in London in which he was associated with Morris Ginsberg, an LSE colleague of Laski, Jennings and Robson. Whether or not there was a direct influence is immaterial: Elias's and Robson's two contemporaneous works on the nature of civilization both centre on the processes by which private acts come to be invested with public significance. This issue exercised an intellectual fascination in the interwar years; thus we find another great contemporary, A. H. M. Jones, charting (with no romantic blurring of the vision) the reversal of this process – the political *decline* of the cities of classical antiquity and the degeneration of their public spirit into squalid individualism – in his *The Greek City from Alexander to Justinian* (1938).

The concern to explain the process by which civilization came to be constructed from barbarism was of course of special significance in the mid-1930s. For Elias (whose 1978 edition is dedicated to his parents, victims of Nazism) and for Robson, for Laski and for many other Jewish intellectuals, the threat of a reversion to barbarism seemed very real. As Robson puts it, formerly free citizens entered under Nazism into 'a state of peonage the like of which has not been seen in the countries of Western Europe for centuries' (Robson, 1940). *Municipal Progress* was not just a tract in defence of local government against centrally imposed expenditure cuts. It was also a warning of the perils posed by the regressive dictatorships.

It is difficult to share, without experience, the preoccupations of another time. Robson was later to recall his increasing concern, after 1932, with

> the threat to world peace posed by the Japanese invasion of Manchuria, the rise of Hitler to power in Germany, the failure of the Disarmament Conference, the progressive decline of the League of Nations, the Italian invasion of Abyssinia, the Spanish Civil War and the weakness of British and French diplomacy. It is difficult for those who have not lived through those years to share the feeling of impending doom. (Robson, 1971, p. 21)

This sense of impending doom is not absent from *Municipal Progress* itself. During the 1930s Robson rarely lost a chance to sound a warning against the New Barbarism, and to underline the significance of the institutions. This he does, powerfully, in the closing pages of *Municipal Progress*, pointing out that there can be no freedom where there is not local autonomy and that the messy pluralism of local autonomy is an early casualty of Fascism:

> There can be no such thing as local government, in the sense in which it is celebrated in this volume, under a dictatorship. Freedom in the localities implies freedom at the centre. Dictatorship at the centre inevitably involves the arbitrary suppression of local freedom and initiative, the curtailment of local responsibility and opportunity. The events of our own time, no less than the experience of history, bear witness to the truth of that conclusion. (p. 464)

Both Laski and Robson recognized that free institutions had to be founded on strength, and that a weak or deficient system of local government would be eroded from within.

There could be no more appropriate epitaph for this volume – one wishes *they* had used it – than Graham Wallas's dictum that '*freedom implies the chance of continuous initiative*'; for, lacking strong pluralist institutions, 'the ultimate implication of the monistic state in a society so complex as our own is the transference of that freedom from ordinary men to their rulers' (Laski, 1925, p. 567). Here then is the underlying theme of the *Municipal Progress*, a concern with the foundations of freedom, sharply focused by the rise of the dictators, but not unmindful of the ever-present danger of the internal collapse of the habit of local initiative. Here too we find the *Municipal Progress* a text for our own time, for 1985, as for 1935.

Acknowledgement

I am grateful to the staff of the British Library of Political and Economic Science, the Public Record Office and the Modern Record Collection, Warwick University Library for their assistance.

2

Economic Change and the Changing Role of Local Government

DIANE A. DAWSON

> Whatever the future may hold for our economic system, local government is likely to remain firmly established as the most effective instrument of social welfare in our national life.
>
> W. A. Robson in *A Century of Municipal Progress*, p. 464

The essence of the Fabian belief in the contribution of local government to social welfare lies in the development of urban public services. A clean water supply, sewage systems, police, public transport and universal education were the services that made for a civilized society. The large cities, the urban areas at the forefront of nineteenth-century economic expansion, were responsible for providing these services. Since it was claimed that local authorities had halved the death-rate in England, it is not surprising that these authorities were seen as effective instruments to control some of the worst social costs of economic growth and to channel part of growing urban wealth into improved living conditions for all.

Today few people would see local authorities as central to improving social welfare for the population at large. There are no doubt many factors behind this change in perception but three interrelated trends in the development of the economy have been of particular importance: long-term economic growth, urban decline and the assumption by governments of responsibility for macroeconomic management of the economy. In addition to these changes in the economic system, there has in the last few years been a change in the ethos of economic policy that would have been unimaginable to the authors of *Municipal Progress*. The Municipal Corporations Act 1835 created elected local authorities but did not confer powers on them. From 1843 to 1929 the assumption of responsibility for water, gas, transport, education, housing and health services was a piecemeal response to the inadequacy or inefficiency of private enterprise and voluntary organizations in supplying these services. By 1935 it appeared that the debate was over. Public provision was winning the day on both social *and* efficiency grounds. The 1980s has seen a reversal of this view. 'Privatization' is only one example of an approach to the economy that presupposes private enterprise to be more efficient than public, and local authorities are frequently condemned for frustrating the government's attempts to reduce the size of the public sector. The debate and the acrimony has produced a climate of opinion hardly supportive of the idea that local government, as provider of services, is as Robson put it 'the most effective instrument of social welfare in our national life'.

Economic Growth

The heyday of local government was associated with the provision of 'public goods' rather than redistributive services. Public goods are essentially services that benefit all as opposed to redistributive services that reduce the consumption of some people in order to increase that of others. The 1835 Act created the basis for elected local councils at a time when economic change was creating the need for local collective action. Britain had been an urban society before the industrial revolution but the rapid growth of manufacturing led to the phenomenal increase in population of major cities. What was unique about this period of urban development was the density of population and the dependency of the economic base on manufacturing.

Manchester, Liverpool and Birmingham were simultaneously the centres of economic growth and of active, interventionist and innovative local governments. The connection was simple. Increased population density changed activities that were previously private concerns or nuisances into public goods or bads. It is unpleasant when a town of 2,000 inhabitants dump raw sewage into the streets or local river; it is a major social and economic problem if this happens in a city of 1 million. Street lighting and policing are not necessary in sparsely populated areas where most people can be expected to know each other. They become essential in areas of dense, rapidly changing populations. Integral to the provision of many public goods was the operation of trading services: water supply, gas, trams and later electricity. These were essentially natural monopolies where the problems associated with private production led over time to municipalization.

Most of these services are in a fundamental sense just as important to social welfare today as a hundred years ago but this does not arouse admiration for local government. There are several reasons. When many of these services were developed in the late nineteenth century they were innovations and a subject of pride. The building of sewers was as much at the forefront of technology and public debate as the railways. Today they can be taken for granted and would continue to be provided if local authorities disappeared tomorrow. The contribution of public health services to the conquest of death and disease provides a good example. The introduction of these services made a major *perceptible* impact on the death-rate. The councillors of Birmingham could record a fall in the death-rate from 25·2 per thousand to 20·7 over only ten years to set against their local investment in public health and the battle that had to be waged in the town council to obtain agreement for the expenditure (Briggs, 1968, p. 225). In their introduction to *Municipal Progress* the editors state that 'Local government in the past hundred years has halved the death rate and reduced the infantile mortality rate by three-quarters' (p. 11). They had in mind data like that in Figure 2.1, especially the downward course of death from infectious diseases. Modern attempts to understand the various factors behind the decline in mortality rates are not as confident in assigning total responsibility to public health measures. The rising standard of living throughout the nineteenth century, especially the effect on diet, is thought to have been a major contributor. Nevertheless, if local investment in public health had not taken place, there is no doubt that much of the improvement would have been halted and reversed by the deteriorating physical environment of the Victorian city (McKeown and Lowe, 1974, ch. 1). The achievements of the nineteenth century were major ones but they have become today's commonplaces.

Education is in many respects similar to public health. Universal primary and secondary education is now accepted as a fact of life.

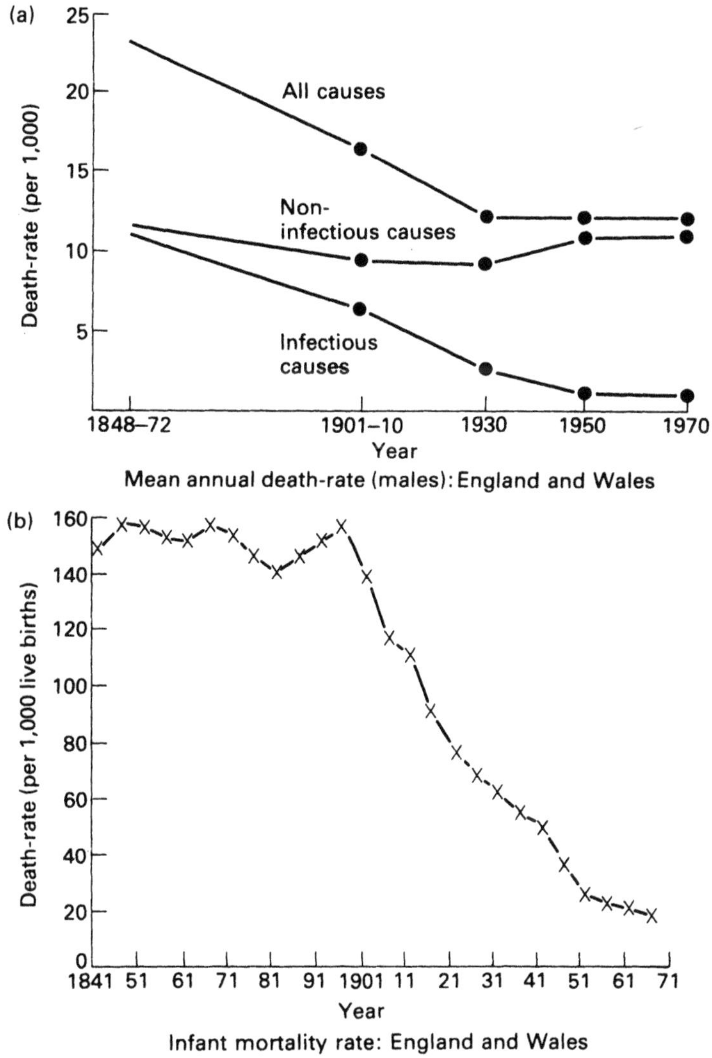

Figure 2.1 *Mortality rates: England and Wales*

Source: McKeown and Lowe (1974)

There are disputes as to how it should be provided, but should local government disappear, there would be no question of reversion to the situation, before the establishment of local school boards, when only around 20 per cent of children had some form of elementary education. As in public health, education was seen as a *local* success because of the long-drawn-out battle to establish the principle of universal education in the first place. The chapter on education in

Municipal Progress points out that the challenge to the 'complacency and to the hesitancy of the Commons came from the municipalities. It was in the towns, especially the new industrial towns of the North, that the supply of schools was most deficient and the state of the children most deplorable' (Smith, 1935, p. 222). The civic societies and pressure groups of Manchester 'had enormous influence on other towns and on national legislation' (ibid., p. 223). In the 1860s 'leadership in the struggle for national education passed [from Manchester] to Birmingham' (ibid., p. 232). Later, local school boards, in spite of government opposition, worked to raise standards above the very elementary education allowed in the 1870 Act (Smellie, 1957, ch. 4). The contribution to social welfare made by the municipal pioneering of universal education is beyond dispute but again the continuation of the system, in some form, can now be taken for granted.

Aside from complacency, there are other reasons why, with higher living standards, local authorities as providers of public services may not be viewed as central to further improvements in the welfare of the population at large. For the present generation the achievement of lower mortality rates, better transport and housing, greater protection from crime, and so on, are increasingly viewed as requiring individual rather than collective action. The private car has released a large proportion of the population from dependence on public transport. The medical profession seems agreed that any further significant reduction in mortality rates depends on individuals' deciding to change their diets and smoke less. Individuals who dislike the environment, housing conditions and crime of the cities can move to smaller towns, and are doing so at an increasing rate. Economic growth has increased the power of a great many people to act on their own initiative.

A related but different point is that the high expectations associated with the initial provision of many public services have faded with experience. Referring back to Figure 2.1, there appears to have been very little change in the overall death-rate since the introduction of the national health service in 1947. The fact that the service reduces the suffering of many who would not in any case die from their illness is of fundamental importance but lacks the spectacular impact of being able to point to a 'concrete result' of public service provision. Similarly, the pioneers of secondary education promoted it as a means of arresting the decline in British industrial competitiveness. Eighty years after the first Act permitting local authorities to levy a penny rate for technical education, decline in competitiveness is still an issue and disillusionment with the failure of the education system to do anything about it is common.

If the role of public services in improving welfare no longer attracts the attention or the hopes reflected in *Municipal Progress*, the issues that today are at the forefront of public debate concern income redistribution. Although Poor Relief was the oldest local government function, local authorities were not responsible for developing modern income maintenance programmes. Pensions and unemployment insurance have from the outset been national programmes. Cash income support is today seen as the main instrument of improving the condition of those most disadvantaged by the modern economic system. The relative emphasis on the redistribution of purchasing power is a consequence of affluence. From the Beveridge Report on, it has been to the central government that one looks for the policies that might make in the second half of the twentieth century a more 'civilized society'.

The prominence of income redistribution in modern economic policy has not only shifted attention from local government to central, but also altered the balance of local authority activities. The slum clearance and community health services that originally were seen as benefiting the community as a whole have been superseded by housing and social services that are designed to be redistributive. In many cases local authorities are trying to fill the gaps or general inadequacies of national cash redistribution programmes. This switch of emphasis from public goods to redistribution has played a part in the postwar tendency to treat local authorities as agents of central government that 'should' act in accordance with the political will of the centre. In 1871 Gladstone was able to say (with some exaggeration) that legislation on local issues did not raise party questions and that one had 'upon the whole, concord and harmony between the two sides of the House' (quoted by Hammond, 1935, p. 39). This statement was simply a reflection of the fact that the issues which at that time interested Parliament, and were therefore the subjects of party conflict, were quite different from the issues that were important to local authorities. When income redistribution is a major source of national political conflict, it is inevitable that local governments pursuing redistributive policies will come into conflict with the centre.

Finally, a rather obvious reason why local authorities are no longer associated in the public mind with essential services is that they are no longer responsible for providing many of them. A significant proportion of *Municipal Progress* is devoted to the history of the acquisition by local authorities of responsibility for services. Writing in 1935, at the culmination of this process, it was not obvious that the next fifty years would require a history of the divestment of local authority responsibility for many of their activities. By 1950 responsibility for trunk roads, gas, electricity and the recently acquired national assistance and hospital services had been transferred to central government control. In 1975 water supply and sewage, the first important successes of local authorities, were transferred to regional water authorities. In some important respects the loss of responsibility for these services was a logical continuation of the processes by which local authorities had acquired them in the first place.

Developments in technology soon transformed what were originally services of local importance into essential components of the national economy. The 1947–8 nationalization of gas and electricity was simply part of a package of measures to nationalize energy, transport and other key sectors of the economy. The fact that most gas and electricity undertakings were already in public ownership made these bills the least contentious of the nationalization measures. The heated parliamentary debates were over the merits of taking privately owned firms into public ownership. There seems to have been relatively little dissent from the view that it was desirable to nationalize gas and electricity in order to secure uniformity of supply throughout the country.

Eventual parliamentary desire for uniformity of services initially developed as *public* services by local authorities is perhaps the most consistent theme in the history of local government in Britain. The 'life-cycle' of local services can be crudely outlined as follows. Conditions in the nineteenth century brought to the attention of a number of outstanding individuals and pressure groups the need for clean water, sewage systems, cheap energy, transport, health and education services. Parliament, however, was either uninterested or highly reluctant to legislate – these were considered to be areas for private philanthropy or private enterprise. The weight of evidence on the importance of these services and on the *unevenness of provision by the private sector* eventually led Parliament to pass *permissive* legislation, allowing local authorities to 'fill the gaps' left by the private sector. In virtually all cases, these local authorities were new *ad hoc* local bodies; even the permissive powers were rarely given to local councils. Local authorities, primarily in the large urban centres, took advantage of such legislation and established public services. The success of these undertakings soon led Parliament to pass *compulsory* legislation to force those authorities that had not acted, or whose performance was considered below the standard emerging elsewhere, to undertake provision of better services. From this stage on, Parliament's main concern becomes that of the *unevenness of local provision*. The often chaotic situation led to the transfer of responsibility from *ad hoc* bodies to local government. Rationalization occurs but there is still diversity of local provision. The solution becomes greater central control of local authorities or transfer of responsibility to the centre.

There was no 'technical' reason why the national health service had to be a nationalized undertaking but a key argument for its establishment was uniformity of service provision. Many of the arguments put forward today for transferring education to the centre rest on the desirability of greater uniformity of service provision. There are objections raised to the diversity of social service provision for the handicapped and pressure on Parliament to secure greater uniformity. This seems to reflect an inbuilt tendency in British politics: once it has been demonstrated that provision of a service is possible and important, centralize it.

Urban Decline

Urban decline is the second major change in our economy that has affected the perceived importance of local authorities. The pattern of nineteenth-century economic development is being reversed with the decline of metropolitan centres and the redistribution of population to small towns and rural areas. The success of local government celebrated in *Municipal Progress* was the success of the new authorities in curbing the excesses of urban industrial development and in aiding economic growth through improvement of the social and economic infrastructure. Today the reverse is true. The large cities are no longer the focal point of economic growth. They are seen as centres of deprivation and decline.

The decline of the great cities began earlier but has accelerated

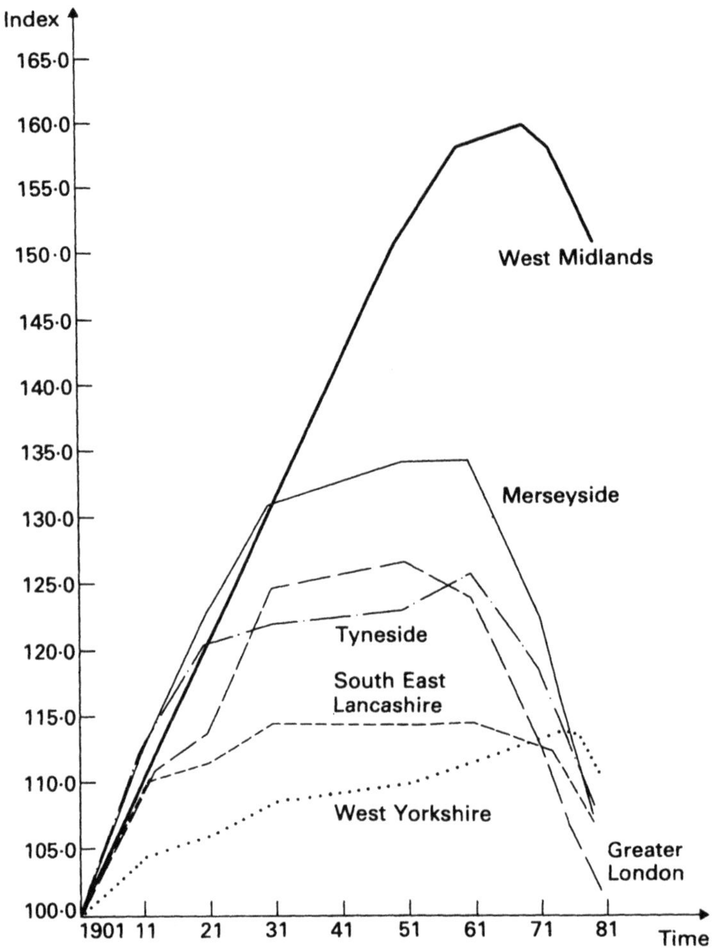

Figure 2.2 *Index of population growth: conurbations*

Conurbation	*Central City*
Greater London	London
Merseyside	Liverpool
South East Lancashire	Manchester
Tyneside	Newcastle
West midlands	Birmingham
West Yorkshire	Leeds and Bradford

Source: McCallum (1980) and 1981 Census Preliminary Report (2)

since the 1950s. At first this represented movement out of the urban core and into adjacent areas. The central city remained the heart of a geographically expanded urban area. The term 'conurbation' came into use in recognition of the fact that the city, as a continuously

Table 2.1 *Manufacturing employment change by type of area 1959–75*

as % of manufacturing employment in each area in 1959

London	−37·8
Conurbations	−15·9
Free-standing cities	+ 4·8
Industrial towns	+16·3
County towns	+28·8
Rural areas	+77·2
Great Britain	− 5·2

Employment change by sector and type of area 1959–75

	as % of total employment in each area in 1959 Agriculture	Mining	Manufacturing	Services[1]	Total
London	−0·3	0	−13·0	+ 1·9	−11·4
Conurbations	−0·3	−2·4	− 7·9	+ 5·9	− 4·7
Free-standing cities	−1·0	−3·6	+ 1·1	+16·0	+12·5
Industrial towns	−1·4	−4·6	+ 6·9	+21·1	+22·0
County towns	−4·6	−0·1	+ 7.5	+15·2	+18·0
Rural areas	−6·5	−1·7	+11·1	+11·4	+14·3
Great Britain	−1·3	−2·2	− 2·2	+10·8	+ 5·1

Notes: [1] includes construction and utilities

Conurbations: The six largest industrial agglomerations, namely Manchester, Tyneside, Clydeside, Merseyside, Birmingham and West Yorkshire

Free-standing cities: Sub-regions dominated by medium-sized industrial cities, e.g. Sheffield and Coventry

Industrial towns: Industrial areas made up mainly of smaller towns, e.g. North-East Lancashire and the Welsh valleys

County towns: Traditionally rural areas containing at least one large town and a moderate amount of industry, e.g. Norfolk (including Norwich), plus coastal resorts, such as Sussex coast

Rural areas: Largely unindustrialized areas, e.g. rural Northumberland, Central Wales

Source: Fothergill and Gudgin (1982)

built-up area and economically integrated labour market, extended well beyond the legal boundaries of the dominant central city. As can be seen in Figure 2.2, even using this broader definition of the city, population decline has been severe in the postwar period. For a time it was assumed that we were simply dealing with a decentralizing population commuting to work but recent research has indicated that the conurbations have been losing jobs more rapidly than population. The 'urban–rural shift' is the term coined to describe this tendency for employment, and manufacturing employment in particular, to gain in smaller towns and rural areas at the expense of the traditional manufacturing cities (Table 2.1).

Explanations for this change in the geographical pattern of economic activity are complex and subjects of dispute. At the highest level of generality it is argued that just as the technology and markets of the industrial revolution led to rural depopulation and urban growth, the evolving post-industrial economy is generating a new, less concentrated pattern of activity. Goddard (1983, p. 25) has suggested that the postwar dispersal of manufacturing 'needs to be viewed in the context of the dying phases of an earlier Kondratieff wave of economic expansion based around electromechanical technologies with firms seeking to remain competitive in outmoded products through improvement in labour productivity'. He goes on to say it is too early to predict the locational implications of the new products generated by the microelectronic revolution.

Most research into urban employment decline has focused on a narrower set of hypotheses. One suggestion was that the decline of manufacturing employment in the large cities was due to industrial structure. The older cities were thought to have a disproportionate share of older industries declining in employment due to Britain's loss of international competitiveness, shifts in consumer demand and the adoption of more modern capital intensive techniques of production. However when attempts have been made to standardize for the mix of industries found in cities, industrial structure would not appear to account for the shift of employment from the large cities to small towns and rural areas. Fothergill and Gudgin (1982, p. 53), summarizing the results of their study of manufacturing employment change between 1959 and 1975, state that their results 'are rather dramatic: industrial structure explains none of the urban–rural shift in manufacturing employment. Indeed, the area with the best industrial structure, London, has actually experienced the worst employment losses.'

Another early explanation was that jobs were simply following people. Rising affluence and declining transport costs were making it possible for a larger number of people to move out of large cities into more pleasant rural areas and small towns (Table 2.2). As the professional and managerial classes were particularly prone to move, newer, growing industries requiring their skills also moved out of the cities. Other hypotheses centring on the movement of industry out of the cities have concentrated on the effects of regional policy during

Table 2.2 *Population change for different categories of districts, 1961–71 and 1971–81*

	Population change: percentage			
	1961–71		1971–81	
England and Wales	5·7		0·5	
Greater London boroughs	− 6·8		−10·1	
Inner London		−13·2		−17·7
Outer London		− 1·8		− 5·0
Metropolitan districts	0·5		− 4·6	
The principal cities[1]		− 8·4		−10·0
Others		5·5		−2·0
Non-metropolitan districts	11·8		5·3	
Large cities[2]		1·4		− 5·1
Smaller cities		2·2		− 3·2
Industrial districts				
(a) Wales and the three northern regions of England		3·7		1·3
(b) Rest of England		12·1		5·0
Districts that include new towns		21·8		15·1
Resort and seaside retirement districts		12·2		4·9
Other urban, mixed urban-rural and more accessible rural districts				
(a) Outside south-east		21·9		8·8
(b) In south-east		22·1		6·7
Remoter, larger rural, districts		9·7		10·3

Notes: [1] Birmingham, Leeds, Liverpool, Manchester, Newcastle and Sheffield
[2] Population over 175,000 in 1971
Source: 1981 Census Preliminary Report

the 1960s and early 1970s and on restrictions on expansion in London and the Midlands. While these policies appear to account for some of the observed change, the predominant influences on the growth (or decline) prospects of different areas appear to be local. Differences in the rates of expansion of *existing* firms and in the birth-rates of *new* firms seem to account for the gloomy economic prospects of the conurbations and the relatively attractive economic future of smaller towns. The term 'urban–rural shift' is in one sense a misnomer; it does not appear that the decline of the cities is due to plants moving or shifting location. In the large cities firms are simply declining more rapidly or expanding less rapidly than their counterparts in smaller towns.

Many of the metropolitan local authorities are now pursuing small-scale local economic development strategies which may, at the end of the day, meet with some success. However there are two points to keep in mind when considering the potential efficiency and success of these programmes. First, it is still not clear *why* the conurbations have a lower birth-rate of new firms and slower growth-rates of existing firms. Until we have a better idea of the causal influences, programmes to alter this situation will inevitably be hit and miss affairs. This applies as much to national strategies for small firms as to local initiatives. Second, there is no reason to believe these programmes will reverse the process of urban decentralization. If they are successful in improving some local employment prospects they will have marginally eased the transition and helped to reduce somewhat the human costs of urban change – an objective worth the pursuit. However there is no reason to believe that government policies, be they national or local, will result in a return to the urban concentrations of the first half of the twentieth century.

The tendency for population and industry to decentralize has been observed in all industrial countries. It is not something peculiar to Britain. This has lent credence to the view that the future economy will not be one dominated by the older cities. Their day has passed. Yet their passing has left, in many cases, the *same* problems we associate with the great activity of local government in the nineteenth century: slum housing, poverty, deprivation, dependence on public transport. Policies to deal with urban deprivation in the nineteenth century could be pursued with two considerable advantages relative to the present. First, the jobs were there. Policy had to deal with the problems traceable to the low wages of the employed – bad housing, sanitation, corruption of food supplies. Second, it could be (and was) argued that programmes to improve the conditions of the working class would strengthen the economic base of the manufacturing city. Public health measures would increase labour productivity, slum clearance would raise property values, cheap workingmen's fares would make a large labour force accessible. Programmes could be presented to the ratepayer, the well-off, as being ultimately in their own interest. A key difference between 1885 and 1985 is that policies to deal with urban deprivation can no longer be sold as making a contribution *to* economic growth. They are now part of a wider and more contentious debate over the degree to which resources should be redistributed from the haves to the have-nots. Ratepayers complain that increased expenditure on these services accelerates decline of the urban economy.

In the 1970s programmes to deal with the inner city problem

began with traditional social services and housing. They floundered on the inability to make any significant difference to deprivation in the face of the evaporating economic base of the Victorian cities. Peter Hall (1981, pp. 12–14), in his introduction to and summary of the studies of the Social Science Research Council (SSRC) on the inner city, captured the distinction between urban deprivation in the context of growing rather than declining urban economies.

> The second half of the nineteenth century, and perhaps above all the 1880s, were distinguished by almost obsessive concern with [the poverty and deprivation of] the inner city .... In the intervening time, it seems, neither analysis nor prescription have shown a really violent shift. There is however one crucial difference. It is that the Victorian city was still a buoyant and growing city. People flooded in to seek their fortune or at least to win a better living than they could find in the countryside ... appalling a place as it might seem by the standards of the late twentieth century, it was – in the words and by the criteria of David Donnison – a Good City. Now ... it no longer seems a magnet; rather, those with the energy or the ability try to escape from it. It does not offer opportunities for its people or good life chances for their children. This is because it has somehow lost the mainspring of successful innovation and of economic creativity. It is, by the touchstones that matter, no longer a Good City. That loss of spirit applies as much to the crucial decisions of the city fathers as to the thousands of small actions by individual entrepreneurs. Civically too, the feeling of enterprise seems to be slipping away.

Hall is perhaps too harsh on local government and its leadership. Local government did not create the economic vitality of the nineteenth-century cities nor can it be expected to halt or reverse the process of deurbanization. The relationship between local government and the local economy is symbiotic. Managing decline, no matter what skills it requires, will never appear as glamorous or exciting as managing growth.

Macroeconomic Policy

The postwar preoccupation of governments with economic management is the third change in the economic environment to diminish the role of local authorities. 'Macroeconomic' issues centring on inflation, unemployment and exchange rates are seen as the issues of the day on which the performance of the party in power at Westminster is judged. The activities of local authorities are discussed in terms of the effects of their expenditure, taxation and employment on central government economic policies and priorities. This change in the policy environment has not only shifted public discussion away from the positive activities of local authorities that so impressed the contributors to *Municipal Progress* but also increased political conflict between the centre and local authorities.

Until the mid-1930s macroeconomic policy consisted of sound money, a balanced budget, managing the national debt and relying on free trade and private enterprise to generate growth and employment. *Municipal Progress* was published just as this economic philosophy was being seriously questioned. In 1936 Keynes published *The General Theory of Employment, Interest and Money*, and the publication in 1944 of the Employment White Paper marked the formal assumption by governments of responsibility for active management of the economy.

The principal technique of demand management, annual manipulation of public expenditure and taxation, produced a radical change in the nature of parliamentary and public debate of local authorities. Between 1835 and 1935 local authority activities were debated in Parliament in the context of specific pieces of legislation: the Private Bills that would authorize major local undertakings, national legislation on education or sanitation permitting or requiring local authority action. Local expenditure *per se*, as long as it was legally incurred, was not a subject of regular scrutiny. Governments were concerned with any costs that would fall on the exchequer but not in the aggregates of local authority spending or revenues.

The financial crisis of 1929–32 was the last clear example of this approach to local government expenditure. The National Government, attempting to balance the budget, sought to cut its share of local authority costs. (The anomalies created and reactions to these cuts are discussed by Young in Chapter 1.) The particular form of the cuts, reduction in wages, was itself a hangover from the policy pursued following the 1925 restoration of the gold standard at the unrealistic prewar parity. To defend the mistaken parity, attempts were made to cut wages in the private sector, a policy that in 1926 led to the General Strike. Wage cuts were seen as the principal step necessary for the gold standard to be reinstated as the *deus ex machina* that would permit government to *withdraw* from the interventionist role in the economy that had been assumed during the war. Checkland (1983, ch. 16) has described this period as 'the last stand' of nineteenth-century economic policy.

Following the Second World War, governments assumed a completely different role in economic affairs. All major economic aggregates, be they private or public, were seen as potentially manageable in the interests of better economic performance. The Keynesian approach to economic management was based on a forecast of the major components of demand in the economy. Local authority expenditure was treated, just as private consumption or central expenditure, as one element of aggregate demand and employment. If the forecast suggested an employment shortfall, a cut in income tax might stimulate consumption, larger capital allocations might increase local authority expenditure and the centre could increase spending on the national health service. If the problem was inflation, public corporations would be told to freeze their prices, local authorities would be given larger grants in order to hold down rates and a prices and incomes policy would try to control wages in the public and private sectors.

The introduction in the 1960s of the Public Expenditure White Papers was the most obvious sign of the tendency to look at local authority expenditure as simply one of the economic aggregates to be manipulated as part of macroeconomic policy. The annual White Paper would begin, setting the context for policy, by indicating the amount by which aggregate demand should grow (or be reduced) and the share within the total that the government wished to see taken by private investment, private consumption and public services. The remainder of the White Paper was taken up by a detailed exposition of the government's plans for public expenditure over the next five years. In these tables local authority expenditure appeared as an undifferentiated part of public expenditure. The government's expenditure plans for, say, education, set out figures for spending on primary teachers (a local responsibility) along with plans for spending on universities (a central responsibility). The commentary on the table would explain why 'the government' planned to spend £*x* million on primary teachers over the period.

This form of presentation had two important effects on the

perceived position of local authorities. First, it made local expenditure a subject of national party political debate whenever the parties differed on national economic strategy. Secondly and more importantly, it gave the impression that the centre determined local expenditure policy, that the totals and the allocation of expenditure between primary teachers, police and road maintenance shown in the White Paper should correspond to actual expenditure. However, the reality had not changed with the form. Local expenditure continued to be determined by local councils on the basis of *their* spending policies, grants anticipated from the centre and local taxation. Hundreds of independently set budgets could, at the end of the day, be analysed to see how much, in aggregate, had been spent on teachers or road maintenance but there was no mechanism linking government expenditure plans with local authority budgets. In the ethos of 'fine-tuning' this situation was unacceptable. During the 1970s the Treasury and governments of both parties were vocal in the need to bring local authority expenditure under control; that is, to enforce its consistency with government plans.

Recent legislation enabling the minister to levy financial penalties on local authorities whose expenditure does not conform to central plans, and the Rates Act 1984 enabling him 'to cap' local rates, are the culmination of the process of integrating local expenditure to national macroeconomic policy. The irony is that these steps have been taken by a government that has abandoned the demand management approach to macroeconomic policy. In many fundamental respects, the current policy environment is much closer to that prevailing during the century 1835–1935 than to the postwar period. Macroeconomic policy is again seen as essentially ensuring sound money, balancing the budget and leaving employment and growth to market forces.

In the nineteenth century this view of macroeconomic policy did not entail control of local expenditure but in the 1980s the government asserts that it is necessary. This change does not reflect any difference in underlying economic theory. While the economic models may be more sophisticated and have acquired the relatively new label of 'monetarist', it remains the case that local expenditure and tax decisions do not affect the money supply. Local authority budgets must, by law, balance and balancing the central budget depends on central expenditure decisions, not local. In 'monetarist' models of the economy, fiscal policy (that is, public expenditure and taxation) does not have a long-term effect on inflation or employment. If all that had happened was a repudiation of Keynesianism, cost-push inflation and other keystones of postwar economic policy, there would be no more reason for the government to take on the task of controlling local authority spending than existed one hundred years ago. (There is a strong economic case for changing the existing system of central grants and local taxation but not for central control of expenditure: see Jackman, 1979.) To understand the present government's desire to acquire the power to control local government expenditure it is necessary to refer back to the third tenet of current economic policy mentioned above: a belief in the efficiency of market forces and private enterprise in particular.

A belief in the efficiency of market forces is strongly associated with most 'monetarist' views of the economy and was even stronger in 1835 when local governments were created. Throughout the first half of the nineteenth century, as the demand for local services developed, Parliament insisted that private companies be given the right to develop them. When, as usually happened, monopoly emerged, the parliamentary response was to set up competing companies in the belief that private competition would ensure efficiency. From the 1850s onward, the policy was slowly and grudgingly abandoned. Robson (1935a, p. 330) thought the evidence had validated public enterprise:

> these enterprises were being run with a degree of ability and efficiency which was embarrassing to profit-making companies in the same line of business, and disconcerting to those who argued that the profit-making incentive was essential to successful management .... The Committee on Industry and Trade, presided over by Sir Arthur Balfour, could hardly be accused of having a predilection in favour of public enterprise – the contrary would be more true. But [in 1928] ... the Committee declared that the general result of their survey was to show that 'the trading activities of the public authorities ... have not lagged behind, and in some cases may have outdistanced, private enterprise in the rate of progress as tested by ordinary criteria'.

Fifty years after this was written, we have a government strongly committed to increasing the role of private enterprise in the economy. In 1980, the first full year of the new government, total

Table 2.3

(a) Government expenditure: 1980 (£ million)

	General government	of which Central government	Local authorities
Current account			
Current expenditure on goods and services	46,673	28,532	18,141
Subsidies	5,308	4,280	1,028
Current grants to the personal sector	27,325	26,200	1,125
Capital account			
Gross domestic fixed capital formation	5,584	1,682	3,902
Increase in the value of stocks	271	271	—
Capital grants to the private sector and public corporations	2,254	1,939	315
Net lending	3,462	2,992	470
Total government expenditure excluding debt interest and non-trading capital consumption	90,877	65,896	24,981

(b) Analysis of current expenditure on goods and services: 1980

	Wages, salaries and other employment costs	Other payments (net)
Total current expenditure on goods and services of which:	29,633	16,929
Central government	15,243	13,436
Local authorities	14,390	3,493

Source: *National Income and Expenditure 1982*. HMSO

public expenditure was equal to 47 per cent of national income. The government entered office with a commitment to reduce overall public expenditure and employment, not as a counter-cyclical measure as in the 1960s and 1970s, but to achieve a permanent change in the balance between public and private sector activity. Three options were open: first, sell off public sector industries and increase use of private contractors; second, cut overall expenditure levels; third, hold expenditure constant so that with growth in the

Table 2.4 *Public sector employment (thousands)*

(a) Employment by sector: 1980		
Central government:		2,327
HM forces	323	
National health service	1,228	
Other	776	
Local authorities:		3,027
Public corporations:		2,038
Private sector:		17,802
(b) *Manpower estimates for 1986/7*		
Central government:		1,935
HM forces	332	
National health service	1,002	
Other	601	
Local authorities:		2,269

Sources: *Economic Trends*, February 1983, and *Fiscal Studies*, May 1984

economy as a whole, there would be a relative decline in the size of the public sector. All three have been key features of policy.

Tables 2.3 and 2.4 give summary statements of public expenditure and employment in a form that highlights the relative importance the new government attached to controlling local authority expenditure. Over half of central government expenditure is on transfer payments – child allowances, pensions, unemployment benefit, supplementary benefit, and so on. Pensions are politically difficult to cut and, during a recession, expenditure on unemployment and supplementary benefits has risen. Only a quarter of central government current expenditure is for the direct employment of people in the public sector and 67 per cent of these are employed in only two services: the armed forces and the national health service. For both services there was a political commitment to expand or protect expenditure. Aside from the national health service, a key characteristic of much central government expenditure is that it does not displace private sector production. Pensions and child benefits are used to purchase goods from the private sector; 57 per cent of defence expenditure is for equipment and services often purchased from the private sector.

In most of these respects, local authority expenditure is the complete opposite of that of the centre; 69 per cent of current expenditure is on direct employment, producing services in the public sector rather than purchases from outside. Local government expenditure is only 38 per cent of central government expenditure but due to the differences in the services for which they are responsible, local authority employment is 130 per cent of central government employment. Transfer payments are a relatively small proportion, 11 per cent, of local expenditure. The only local service the government was committed to expand, the police, represented just 6 per cent of local authority employment and 10 per cent of local current expenditure. For any government committed to a policy of reducing the role of the public sector in the production of goods and services, control of local authority expenditure was of far greater importance than its share in total government expenditure would indicate.

There have been a number of attempts by economists to investigate the question of whether there is any significant tendency for a private producer of a service to be relatively more efficient than a public producer. A recent survey of this work concluded 'that there is no systematic evidence that public enterprises are less cost effective than private firms', but qualified this conclusion with an important reservation on data availability: 'at present observations are often available only for selected years with long time series conspicuous by their absence. This is not a problem unique to this area of applied economics but the requirement of comparing the two institutional forms on a similar product basis certainly makes life more difficult' (Millward and Parker, 1983, pp. 257–8).

Whatever the economic evidence, there is no doubt that the government plans to continue to attempt to reduce public spending and local authority expenditure in particular. Judging from figures in the 1984 Public Expenditure White Paper, the government is hoping that by 1987 it will have found the means to reduce local authority employment by 25 per cent of its 1980 level. While the policy *per se* would inevitably have its opponents, it has been the interaction of this economic strategy and urban decline that has produced the most serious clashes between central and local government. It is not by accident that the metropolitan counties and districts have been the hardest hit and the most vocal opponents. The problem can be illustrated by reference to two major services where the government states that it has *not* cut real expenditure: education and health.

The government's plans held constant real expenditure *per pupil*. Since the number of school-age children has been falling, this would require a reduction in the number of teachers employed nationally.

However, the older cities are losing population to the rest of the country at a rapid rate so their fall in pupil numbers is even greater than the national average. Their employment cuts, to hold expenditure per pupil constant, would have to be proportionately greater than elsewhere. The same is true of the national health service. Funds are allocated on a weighted population basis. Holding the total constant in real terms means there must be cuts in facilities and employment in the cities losing population as funds are reallocated to the areas of growing population.

There is evidence to suggest that as the conurbations were losing private sector employment to the rest of the country, a major factor in preventing an even greater rise in inner city unemployment was the growth of public sector employment. Now this safety valve is being closed and in many of the older cities current economic policy is seen as accelerating their already serious economic and employment decline.

Conclusions

In a fundamental sense, events of the last fifty years have not undermined the conclusion in *Municipal Progress* that the services which in 1935 were provided by local authorities are and will remain central to social welfare. What has changed is that local authorities are no longer seen as the necessary 'instruments' for delivering those services. Some services have already moved to the centre. Experience of the last few years makes it easy to imagine a future where local authorities have been converted into bodies similar to district and regional health authorities – 100 per cent centrally financed and responsible for their policies and priorities to central government. There could be a reversion to fragmentation with one local board administering education for the Department of Education and Science (DES) and another running environmental services reporting to the Department of the Environment (DoE). Whatever the organization, the major services of 1935 will continue as the bedrock upon which other services both public and private rest.

Most of the contributors to *Municipal Progress* would not be surprised by pressure towards uniformity and centralization of services. In places their ambivalence is explicitly acknowledged. In his chapter on education, after commenting on the role of 'variety, elasticity and experiment' in the development of education service, Professor Smith (pp. 237–8) adds:

> The maintenance of variety within the system will be more difficult. Unification of administration tends towards uniformity .... Perhaps we owe some of this elasticity to the variety and even chaos of the nineteenth century; if so, the reformers whose task it is to keep alive the spirit of enterprise will have a harder task in the next generation, for the machine grows increasingly efficient.

While centralization of public services may have been anticipated fifty years ago, urban decline was not. Today there is no necessary connection between local government and urbanization; a rural community can be expected to take as much interest in its local schools and other public services as an urban one. But historically it *was* the large city that took the lead in service development and the rural counties that were forced, by Parliament, to follow that lead. The implicit belief of the contributors to *Municipal Progress* was that continued urbanization would further increase the sphere for communal satisfaction of social wants. This has not happened.

Urbanization *per se* was a critical factor in service development but just as important was the rise in living standards that accompanied the growth of Victorian cities. It was the wealth of the cities as well as their problems and civic pride that enabled them to pioneer public services, converting the luxuries of one age into the necessities of the next (Robson, 1935a, p. 300). In the postwar period the link between urbanization and economic growth has become more tenuous. The famous cities now need financial aid to deal with the problems of urban decline. Changes in patterns of production and consumption have made improved living standards more likely outside the city than within it. Postwar economic growth has enabled many people to concentrate on satisfying their private wants. Today major improvements in social welfare hinge on national programmes for the redistribution of private purchasing power. During the period 1835–1935 economic growth took a form that gave local government a prominent place in social improvement. Laski, Jennings and Robson did not foresee that as living standards continued to rise, the programmes central to further improvement would be national ones.

Perhaps the most profound difference that fifty years has made to the view that local government is 'the most effective instrument of social welfare' has been the postwar experience of central government macroeconomic policy. Local authorities are now simply expenditure aggregates that do or do not behave in a way that is consistent with government attempts to deal with unemployment, inflation and economic growth. If it is in the power of national governments to secure these ends, then obviously it is the centre, not local government, that is the primary instrument of social welfare.

In fact, governments cannot effectively control all these variables but they adopt policies and rhetoric that suggest that they can. No matter what the complexion of the government, there is a common belief that national economic performance can be improved if central government policies are pursued. In the nineteenth century growth and employment seemed to be based in urban activities and local councils could be seen to have a constructive role to play in the process. It was Keynesianism that changed the perception of the role of the national government in the economy and even a government rejecting that framework has inherited the mantle of responsibility.

3

Structure, Centralization and the Position of Local Government

ALAN ALEXANDER

> The first and most pressing matter which calls for attention is the structure of our local government system . . . It is reasonably certain that unless a major adaptation of the structure is effected before we reach the second half of the twentieth century, an alternative solution will be sought through the establishment of boards or commissions set up to conduct local services and entirely divorced from existing municipal institutions
>
> W. A. Robson in *A Century of Municipal Progress*, pp. 456–8

Of all the contributors to the *Municipal Progress*, William Robson was most concerned with questions of structure. His warning in Chapter 19 that the continued health and strength of local government would depend on a comprehensive and, by implication, radical redrawing of its boundaries and redistribution of its functions was just one of the many occasions on which he returned to the theme over a period of thirty-five years (Robson, 1931a, 1935b, 1948, 1966). Robson was right to see the problem as urgent in 1935. Looking back from 1985, what needs to be explained is why it took so long for government to accept that something had to be done.

The Acceptance of *Immobilisme*, 1935–45

In the ten years or so after the publication of *Municipal Progress* support for the view that the local government structure was passing from the obsolescent to the obsolete came from political, official and intellectual sources. In 1940, for example, the Report of the Royal Commission on the Distribution of the Industrial Population (Barlow, 1940, para. 151) had observed:

> The network of local government areas has . . . retained the main outlines of the structure established in the last century in conditions vastly different from those of today . . . The important industrial towns have long outgrown their boundaries as local government units.

Professor Abercrombie made much the same point in the early 1940s in his series of reports on the planning needs of Greater London, which were influenced by the conviction (shared by the department which commissioned them) that London could not be planned adequately within the existing framework of local government areas and, indeed, that that framework had become a positive hindrance to comprehensive medium- to long-term planning.

Barlow and Abercrombie, in common with a small number of academic students who thought and wrote in the same vein as Robson and his *Municipal Progress* co-authors, failed however to be clearly explicit about the major obstacle to reform. The quotation from Barlow gives a clue in its reference to the clear inappropriateness in modern circumstances of the boundaries of the 'important industrial towns'. For the structure of local government, imposed, as Robson (p. 456) colourfully put it, by three 'volcanic eruptions . . . in 1835, 1888 and 1894' upon 'areas . . . formed by a gradual process of sedimentation', was based upon an historically derived but functionally indefensible distinction between the local government of urban and non-urban areas. The reform of the municipal corporations in 1835, the extension to county councils in 1888 of the principles of representative government, and the completion of the structure by the creation of the district councils in 1894 and of the London metropolitan boroughs in 1899 can be viewed in retrospect as a classic example of incremental institutional change producing a result that was both irrational and functionally ineffective. Robson and others saw this in the 1930s. Why then did it take almost thirty years before major structural change was begun and nearly forty until it was completed?

The most important reason was that the outmoded distinction between the urban and the non-urban areas was perpetuated in the institutional structure. The upper tier of the local government structure as it emerged from the nineteenth century comprised counties and county boroughs, while the lower tier comprised rural and urban districts (and in London their virtual equivalents, the metropolitan boroughs). The problem was not merely that in many parts of the country the distinction between urban and non-urban existed only in the institutional structure; more importantly, the separate institutions were locked into a conflict which was imposed by the structure and which acted as a major inhibitor of structural change. This problem is best illustrated at the level of the major units of the unreformed system – the counties and the county boroughs. For it was here that the misfit between demography and institutional structure was most obvious.

The boundaries of the county boroughs were set in 1888, when sixty-one towns were excluded from the jurisdiction of the new councils set up under the County Councils Act. Within a very few years, the process of suburbanization had made a nonsense of these boundaries (Young and Garside, 1982) and the Local Government Board and Parliament were continually forced to consider applications from county boroughs for extensions to their territory, as well as applications from urban districts and municipal boroughs for promotion to county borough status. Boundary extensions were always opposed by the county councils, for they realized that they were engaged in a zero-sum game – in which what one side gains, the other must lose – with the county boroughs over territory, tax base and, in the case of applications to become county boroughs, status. Parliament, when it sanctioned changes to boundaries, was inevitably pushed in the direction of a judgement of Solomon, a decision between antagonists which both limited change and confirmed the permanence of the basic institutional structure. Thus, the conflict that was built into the local government system in the nineteenth century created *immobilisme* in the twentieth, to the extent that in the 1920s two Royal Commissions (Ullswater in London and Onslow elsewhere) found it impossible to agree on any recommendations for fundamental changes to the system.

The adversary relationship at the local level between the classes of authority was, inevitably, reflected in their national representation. The county boroughs (and, confusingly, the municipal or non-county boroughs which functionally had more in common with the urban districts) were represented by the Association of Municipal

Corporations (AMC) and the county councils by the County Councils Association (CCA). The lower-tier authorities established in 1894 were represented by the Urban District Councils Association and the Rural District Councils Association. Whether in terms of representation on investigating bodies, such as royal commissions, or in responding to particular government initiatives or to specific applications for changes in boundaries or functions, these associations fought their own corners, defended the institutional and territorial integrity or ambition of their members, and paid little regard to the overall structural and functional coherence of the system. Robson (1948, p. 153) was withering about these local government representative bodies, seeing in them 'the worst aspects of vested municipal interests' and condemning them for failing to show 'any of the qualities which might have been expected from civic bodies entrusted with the welfare of local communities', but he was less than fair, as zealots for reform often are. The associations could not, given the structure in which they worked and from whence they sprung, act in any other way. They existed to protect and defend the perceived interests of their member councils, so the zero-sum game that was played locally between expansionist towns and conservative counties was reflected in the attitudes struck and the representations made whenever and however the structure of the local government system reached the national agenda.

For twenty years after the publication of *Municipal Progress*, the central departments responsible for local government seem to have accepted the intractability of the existing structure despite increasingly frequent admissions of its obsolescence. In 1941, for example, the wartime coalition government, perhaps stimulated by the need to construct an entirely new system of regional administration for emergency planning purposes, appointed a committee to look at the postwar needs of the local government system. In 1944 the House of Commons was told that there would need to be changes in 'status, boundaries and areas', which suggested the possibility of structural modernization, but that these changes would be 'within the general framework of the county and county borough system', a qualification open to two quite different interpretations (House of Commons, 1944). On one reading, it indicated an acceptance by ministers, and more importantly by the central departments concerned, that since consensus within the existing local government system was not available, any government initiative would have to be marginal, and confined, in a vivid image used much later by the Wheatley Commission, to 'surface patching'. On another (and this view was given some credibility by the work of the Local Government Boundary Commission (LGBC) between 1946 and 1949) the statement to Parliament suggested that while a thorough-going and radical recasting of the system was not immediately in prospect, the possibility of major reform by cumulation existed, as a government-sponsored commission, with the approval of the local authority associations, was to examine and make recommendations for change where the misfit between area and function had become most serious. Whichever view is taken, ministerial initiative was hedged in by a search for consensus that gave local government a virtual veto over government proposals.

The Search for Agreement, 1945–57

By the time of the publication in January 1945 of the Coalition Government's White Paper, *Local Government in England and Wales during the Period of Reconstruction* (Cmd 6579, 1945), the Labour Party, the National Association of Local Government Officers and many planners were convinced of the need for a major structural reorganization. The LGBC suggested by the White Paper seemed consistent with the more far-reaching of the two interpretations of the 1944 statement: new and enlarged county boroughs might be created; and declining ones might be given a new status more in keeping with their modern position. In the text of the White Paper, however, there were statements of policy on the nature of change which made the first interpretation more credible. The terms of reference of the LGBC, reflecting the White Paper, excluded it from considering *functions*: it could not, therefore, seek to match area, population, resources and functions. It could not look at all at the area of the London County Council or recommend the creation of new county boroughs in Middlesex. Instead, and elevating to the status of policy the practice of 'ad-hocery' warned against by Robson, the government's view was that co-ordination of functions 'should be sought by the establishment . . . of Joint Boards and Joint Committees' (p. 8). This exclusion of London from the remit of the LGBC is evidence also of the piecemeal approach of governments to the local government system, for it arose from the fact that there was already in existence a Departmental Committee under the chairmanship of Lord Reading, examining the problems of local government in London.

Despite these obvious limitations on its activities, some commentators believed that the LGBC offered the prospect of cumulative and progressive structural reform (Lipman, 1949, p. 249). The commission itself thought differently and in 1947 it complained in its annual report of its lack of jurisdiction over functions: 'Our present powers and instructions do not permit the formation of local government units as effective and convenient as in our opinion they should be' (LGBC, 1947, p. 2). Also, in a direct appeal to the responsible minister, Aneurin Bevan, the report quoted with approval a comment made by him in a parliamentary committee to the effect that 'it is nonsense to talk about functions and boundaries separately' (ibid.).

Bevan's view of the local government system is a little difficult to determine, and indeed the whole attitude of the postwar Labour government seems rather confused. For a radical and *dirigiste* government the effectiveness of local institutions was clearly important, and as Gyford shows elsewhere in this volume, Labour in 1945 was the first national party explicitly to link the two stages of the policy process: national goal-setting and local implementation. In 1946 a junior minister, echoing an undated policy document published some time in the prior two years (Labour Party, n.d.) told a Labour Party regional conference that he was in favour of a system of regional councils supported by most-purpose local authorities. In 1948–9, Bevan prepared, but did not press in Cabinet, a radical plan to abolish the existing system and replace it with 240 new all-purpose authorities (Foot, 1975, pp. 263–4). In late 1947 he made the statement about boundaries and functions quoted back at him by the LGBC (House of Commons, 1947). In March 1949 he told the House of Commons that 'it will not be practicable to introduce comprehensive legislation on local government reconstruction in the near future' (House of Commons, 1949). And in 1949, the LGBC was abolished.

It is likely that Bevan's inconsistency with regard to local government arose from an institutional factor that, together with the adversary relationship of the classes of local authority, helps to explain the structural *immobilisme* of the years between 1888 and 1957. Britain has never had a Ministry of Local Government, and there has not therefore been a Cabinet minister whose primary concern has been the local government system. Even in the days of the Local Government Board, the responsible department between 1872 and 1919, the dominance of the Poor Law tradition made it all but impossible for ministerial initiatives to be taken on the structure of municipal institutions. Bevan was Minister of Health and, as such, he was pretty fully occupied, in the period to which the above chronology refers, with the establishment of the national health service, an institutional innovation which, incidentally, had the effect of reducing rather than enhancing the responsibilities of elected local institutions. Similarly after 1951, when Harold Macmillan was Minister of Housing and Local Government, the emphasis was on the housing programme, largely executed by and facilitated through the housing and planning powers of local authorities, rather than upon the continuing and worsening institutional obsolescence of the local government system.

Despite the LGBC's summary abolition, it had, by the very act of questioning its remit, moved the issue of major institutional change in the local government system a little further up the agenda. By proposing the 'new county boroughs' and the new single-tier counties in continuous urban areas, the commission became the first official body to say that radical change outside the county/county borough system was both practicable and desirable. Further, by emphasizing the limitations placed on its task by the exclusion of London and Middlesex from its terms of reference, the commission was able, in its report for 1947, to underline the particular urgency of the problem of local government in the Greater London area. It asked whether it could make recommendations for Greater London outside the LCC area and Middlesex on the assumption that these areas were not to be examined at all:

> It would, in our view, be wasteful in money and manpower if, after we had made our proposals and the elaborate machinery to give effect to them had been put in motion, it was then decided that London and Greater London should be reviewed as a whole. (LGBC, 1947, p. 40)

Once again, the Commission quoted a remark in the Commons by Bevan, referring to the Departmental Committee on London Government, that the problem of governing the capital 'cannot satisfactorily be divorced from that of Greater London' (ibid.).

In the twenty years after the publication of *Municipal Progress* there was no shortage of expressions of concern about and accurate diagnoses of, the ills of the local government system. There is, moreover, evidence that ministers knew what had to be done. But as the consultations preceding the establishment in 1957 of the Royal Commission on Local Government in Greater London and the creation in 1958 of the Local Government Commissions show, the search for agreement in a deeply divided local government community continued to inhibit the enthusiasm even of ministers who took an interest in the local government side of the ministry's work.

This search for agreement where none was possible was particularly assiduous between 1954 and 1956 when, before publishing a statement of government policy, the minister, Duncan Sandys, negotiated with the local authority associations an agreed 'Statement of Principles' upon which reform of the local government system would be based. The tone of the agreement, which was published as an appendix to the White Paper, *Areas and Status of Local Authorities in England and Wales* (Cmd 9831, 1956) was not radical. Rather, it was in the mainstream of the search for an agreed solution and, in fact, it represents the last occasion when the local authority associations were able to exercise their veto on major structural change. Nothing had been done about the decay of the local government structure between 1949 and 1956. The Herbert Commission, in its Report on Local Government in Greater London (1960), listed five basic reasons for its own establishment, the fifth of which applied with almost equal force to the system outside London:

> The unsettling effect of the situation which prevailed, with varying degrees of intensity, since 1945. The need for reorganisation has been in the air throughout this period. (Herbert, 1960, p. 18, para. 70)

That uncertainty had, moreover, been highlighted by changes to what might be called the functional profile of local government. It has already been mentioned that the establishment of the national health service, by removing from local authorities the management of hospitals, had reduced the range of functions. Similarly, the nationalization of the public utilities of gas and electricity production and supply constituted a major limitation on the trading activities of local government, while at the same time creating new and powerful 'statutory undertakers' with which councils would have to deal, especially in the fields of development and land-use planning, where the powers and duties of local authorities were being radically altered. Also, the sheer size of the task of postwar reconstruction had ensured that in housebuilding, county boroughs and urban districts were no longer their own masters in the sense that for most of the interwar period they had been, when, where they had any housing programme at all, they planned and largely financed it with little direction from the centre. After 1945 local authorities became more the agents of a national housing policy than autonomous providers for their local communities.

A similar process affected education under the Education Act 1944, although here, as the continuing debate over the form and content of secondary education suggests, the nature of central control was such as to leave education authorities a degree of autonomy within the context of the obligation to submit, for the approval of the Ministry of Education, development plans for all stages of education provision within their areas. Central control was also exercised through the process of capital expenditure approval, the continuation of the national negotiation of teachers' salaries and the approved quota system which governed the number of teachers that education authorities might employ.

Finally, the National Assistance Act 1948, the first clause of which indicated the gradualness of change in the local government system by providing that 'the Poor Law of 1601 is hereby repealed', at last removed from local government that responsibility by which, especially at the lower tier, its nineteenth-century structure had been so deeply conditioned – the relief of the poor. Local authorities retained and acquired related functions in public health, environmental health and social welfare, but there was a clear anachronism in the survival of an institutional structure whose long historical continuity arose from its responsibility for a service it no longer provided.

The appointment in 1957 of the Herbert Commission on Local Government in Greater London marks a turning-point in the modern history of British local government. The decision to look comprehensively at the metropolitan conurbation, which contained around a hundred local authorities, each potentially a centre of opposition to radical change, may be seen as the point at which central government asserted its constitutional ascendancy over local government in a way that extended beyond the supervision and control of individual functions and services. The search for consensus had been based at least partly on a recognition of a form of constitutional autonomy attaching to local authorities, especially the ancient municipal corporations, but extending also to other authorities which were, in a sense, carried along with them. In a constitutional system based upon custom and precedent, the status and continuity of local institutions protected them against total domination from the centre. Thus while they were subject to increasingly detailed control from the middle of the nineteenth century, their institutional integrity remained largely inviolate. In this context, it is significant that the only systematic attempt at modernization before 1935 affected only the county districts – the most recently created local authorities and a class of institution that did not raise emotive issues of status and dignity. Michael Elliott (1983, pp. 44–5) has recently pointed out that the significance of the Local Government Act 1972, from a constitutional point of view, lies in its explicit rejection of the ancient notion that municipalities drew some of their legitimacy from powers and status granted by royal charter: even the Municipal Corporations Act 1835 had not interfered with the constitutional status and corporate autonomy of the borough councils. The events leading to the reform of local government in Greater London, starting with the appointment of the Herbert Commission, indicated that the constitutional position of local government would undergo a profound change.

Structural Change, 1957–72

The reform of local government in the metropolis revealed, however, that even when the circumstances are most propitious for the assertion of the supremacy of national government, the absence of any extra-parliamentary method of institutional change may render the process discontinuous and the outcome less than rational. The parliamentary position of the Conservative government that appointed the Herbert Commission was greatly strengthened after the 1959 general election. Following the receipt of the commission's unanimous report, the government was in a strong position to get its own way in producing a new scheme of government for London, one which was drawn from the conclusions of an impartial royal commission but would also maximize the Conservative Party's electoral advantage in London. The route between intent and achievement, however, lies through Parliament. Threatened institutions are, therefore, guaranteed by the legislative process a forum in which to defend their existence and, if they can in turn threaten either the government's majority or its parliamentary timetable, they may extract concessions. Such concessions may ensure the passage of the Bill while compromising its internal coherence.

In the case of London government, the most powerful of the threatened authorities was the London County Council (LCC). The abolition of the LCC was seen as in the Conservatives' political interest for it had been Labour-controlled continuously since 1935. By submerging the LCC in a Greater London Council (GLC) and by distributing its powers among that council and the proposed London borough councils, the new system would, at a stroke, fragment Labour power in inner London, and in the government of the metropolitan conurbation, bring in the outer London suburbs as a counterweight to the strength of Labour in the inner city. The GLC was intended to be almost as much a Conservative stronghold as the LCC had been a Labour one.

As soon as the government announced its acceptance of the main recommendations of the Herbert Commission the LCC, like all the authorities adversely affected by the proposals, began to fight back. In the event, the result of various parliamentary concessions was to preserve the integrity of the LCC education service by the creation of the Inner London Education Authority (ILEA), technically a special subcommittee of the GLC but in practice a single-purpose, indirectly elected local authority, providing education to the twelve inner London boroughs created in the territory of the LCC. The establishment of ILEA was a response to internal Ministry of Education pressure but it has worked to Labour's advantage in inner London (Rhodes, 1970). At the same time, the government bowed to Conservative pressure on the outer boundaries of the GLC area. This led to the exclusion of some suburbs in Surrey, Hertfordshire, Essex and Kent and so made Conservative control of the GLC less certain, and in the event the new council was Conservative controlled only in the periods 1967–73 and 1977–81. The immediate result, as a consequence of the interplay of the search for partisan advantage and the process of parliamentary bargaining, was a structure of government in the metropolis which was more complex than originally intended and which, by the creation of ILEA, ran counter to the historic trend in favour of directly elected multi-purpose authorities. Also, the confusion of the search for efficient and convenient units of local government with the pursuit of partisan advantage and the need to appease party opinion had produced a GLC which might possibly succeed in providing the kind of strategic planning system generally agreed to be necessary but which would do so within boundaries which coincided with neither the limits of suburbia nor the long-established area of the Metropolitan Police.

The GLC, ILEA and the new London borough councils took over in 1965, having been elected in 1964 to serve as 'shadow' authorities preparing the transition to the new system. During the transition, services continued to be provided and even the most antagonistic of the existing councils was unable to frustrate the transfer of power. The government made compensation available to any staff whose jobs might disappear and guaranteed preferential access to posts in the new authorities. The whole exercise demonstrated that radical institutional change was practicable, that it could be achieved despite the opposition of existing institutions, that the device of the impartial royal commission could effectively counter vested interests and local power and that the pursuit of party advantage could be an integral part of the process of reform. Reorganization was completed in London just as it became clear that none of these conclusions could be drawn from the piecemeal reform process under way in the rest of England and Wales.

In 1958, the year following the appointment of Herbert Commission, the government decided, with the agreement of the local authority associations, to appoint Local Government Commisions (LGC) for England and Wales. The commissions would, in general, be able to recommend changes to the boundaries of counties and county boroughs. Only in the Special Review Areas of Tyneside, West Yorkshire, South East Lancashire, Merseyside and the West Midlands would the Commission for England be able to look at the distribution of functions between the tiers of local government. The appointment of the commissions reflected the continuation of the departmental view, personified by Dame Evelyn Sharp, the Permanent Secretary to the Ministry of Housing and Local Government from 1955 to 1966, that the influence of the existing authorities and their associations was such as to make it impossible to consider radical, comprehensive change. The 1956 White Paper (Cmd 9831) had said that although the government did not believe 'that no changes should be proposed . . . except with the agreement of the local authorities', it accepted that 'their willing co-operation can be of great value in carrying through any reorganisation' (para. 14).

The disadvantages of the LGC approach to reform were the mirror images of the advantages that could be attributed to the Royal Commission–White Paper–consultation–legislation approach employed in London. Radical change, except perhaps in the limited confines of the Special Review Areas, was impracticable; the process of local investigation, report and ministerial decision gave ample scope for existing authorities to frustrate change; and the fact that final decisions were taken by the minister made it difficult to legitimize the search for party advantage in particular localities by placing it in the wider context of comprehensive institutional change. As the Minister for Housing and Local Government in the Labour Government elected in 1964, Richard Crossman, told his diary 'it meant accepting the war between the county councils and the county borough authorities as endemic in our national life' (Crossman, 1975, p. 65). That diary entry was made a month after Crossman became minister, but the resilience of the departmental view is apparent from the fact that it was not until almost a year later that Crossman announced that the LGC would be wound up and replaced by a royal commission. In the intervening months, he had faithfully spoken to the departmental brief, telling various conferences that radical reform was a non-starter and that 'the county councils are always going to be with us'.

The royal commission device, by 1966 when the royal commissions on Local Government in England (the Redcliffe-Maud Commission) and Scotland (the Wheatley Commission) were appointed, had demonstrated its usefulness in breaking the log-jam of vested interests standing in the way of reform. For although a royal commission might, in Crossman's words (Crossman, 1975, p. 331) be 'a committee of enquiry with very great authority' it had, for national politicians, other important characteristics. First, it was possible to co-opt the vested interests to the process of investigation by giving them representation on the commission; second, although the commission might, through the balance of its membership, achieve an appearance of impartiality, the need for a government response to its recommendations meant that ministers would control the vital stages of the reform process; and, third, the combination of impartiality, representativeness and authoritativeness provided an effective cover for the pursuit of partisan advantage which, by the 1960s, had become an important factor in the reform process. National parties were more aware than they had been thirty years before of the link, alluded to earlier, between national goal-setting and local implementation. That awareness, with its inherent pressure towards centralization, was a powerful stimulus to change, not only in the shape of the local government system, but also in the place and position of local government in the political system at large. That pressure, as will be argued below, was in addition to a long-standing tendency towards centralization in the relationship between the centre and the localities.

The election of the Labour government in 1964 ushered in a period when institutional reform was fashionable and when there was an intellectual and political orthodoxy in favour of pursuing imagined economies of scale. The broad critique of the local government system which informed the decision to set up the royal commissions was in three parts. First, there were too many local authorities and most of them were too small to be effective or efficient. Second, the boundaries even of those authorities which were big enough to be effective service providers were obsolete, and modernization by the LGC process was cumbersome, bureaucratic and slow. Some new boundaries might be obsolete before others had been reviewed at all. It was also uncertain in that the need for ministerial approval of any recommended changes inevitably meant that changes could be rejected for reasons unconnected with the coherence of the local government system. Third, these first two factors combined to ensure that, in the words of Redcliffe-Maud (1969, para. 6(i)), 'Local government areas no longer correspond[ed] to the pattern of life and work in modern England'.

There was little dispute by 1966 that something had to be done about the local government system. The process of change was, however, more complex, discontinuous and irrational than had been the case in London. The need to get reform through Parliament in the form of partisan, public legislation ensured that the attitudes of the major political parties, which are considered below, would condition the nature of reorganization. Specifically, the progress from royal commission to reform was more tortuous outside London because a general election, and a change of government, intervened between report and reform, allowing both parties publicly to reveal their preferences and prejudices, and ensuring that the government which appointed, and to some extent packed, the royal commissions would not determine the fate of their recommendations. The process was also complicated by another factor, one which reduced the agenda-setting and policy-forming effect of the royal commission as noted in the case of the Herbert Commission: the Redcliffe-Maud Commission (unlike the Wheatley Commission) failed to deliver a unanimous report and its members, especially its chairman, were less than robust in defence of the majority recommendations. Finally, Redcliffe-Maud was, in its justifications for recommending unitary authorities serving between 250,000 and 1 million people, somewhat unconvincing. Its major research study on the optimum, and minimum size of authority required for the efficient provision of particular services was, as the author of the Memorandum of Dissent suggested, an insubstantial basis upon which to build its preferred scheme. The study had been commissioned at a very late stage in the commission's work, apparently in order to justify a conclusion already reached. It is difficult not to conclude that Redcliffe-Maud was a victim of the prevailing orthodoxy, noted earlier, that big was beautiful.

The outcome of the reform process was the creation of a new system of local government which greatly reduced the number and greatly increased the average size of authorities. By creating two-tier structures everywhere, the reorganization ensured that the 'most fatal defect' (Redcliffe-Maud, 1969, para. 85) of the old system, the division of town and country, would be replaced by an equally debilitating defect, the failure to recognize the predominance and the particular problems, at least outside those conurban areas that became metropolitan counties, of major urban centres. And by ensuring that the new classes of authority would be as mutually antagonistic as the old ones had been, it ensured that local government would be ill-equipped to resist the rapid increase in the pressure for centralization.

The Political Dynamics of Reform

In the relationship between central and local government, there has always been an inherent tendency towards centralization and it is important to be clear about its character. This tendency requires us to recognize that the position of local government within the hierarchy of British governing and policy-making institutions has deteriorated considerably in the past half century and that the decline has been rapid in the last ten years. This is not a temporary decline in the fortunes of subnational institutions, nor is it a short period of centralization between longer periods of stability.

The tendency towards centralization had until recently two sources. First, there is at the centre an urge for power over the localities. This is essentially a *political* urge. It does not arise from the subordinate legal position of local government, for there is no reason why the principle of *ultra vires* need imply centralization. It would be possible to maintain the supremacy of national government, which is argued to be necessary to ensure minimum and comparable standards across the country, without progressively limiting the scope for local initiative. The inexorable tendency towards centralization arises, on the political side, from the habit of national politicians in setting national goals that can be reached only through the activities of local institutions. Put in party political and electoral terms, this means that national politicians make promises that can be kept only with the co-operation of local authorities. Thus, to some extent, it is inevitable that local authorities will be the administrative and implementing arms of central policy-makers. Three examples from recent years illustrate the point. The Labour Party commitment to the comprehensive reorganization of secondary education, the Conservative manifesto promise to give council tenants a right to buy their homes, and the post-1979 Conservative government's target of reductions in local public expenditure all relied for their achievement on the co-operation, compliance or coerced acquiescence of individual local authorities.

There are, however, discernible differences in the quality of the goals set by national politicians. Although the effect of any nationally set goal must be to diminish, or at least to limit, the autonomy of local government, the perception of that effect at local level will be conditioned by whether the target entails the expansion or the contraction of local services. When national government can provide additional resources, any re-gearing at local level is lubricated by money that need not be raised locally in the form of an increase in that most palpable and unpopular of taxes, the domestic rate. But, even in this event, the capacity of councils to resist both the increase in central power and the 'bribery' that comes with it may sometimes lead, as with the Labour government's education policy and the Conservatives' commitment to council house sales, to specific legislation which permanently reduces the discretionary power of local authorities.

In the context of the shape and nature of the local government system in general, the urge to centralize power manifests itself in attempts to influence the structure in ways that maximize at local level the advantage of the party that happens to be in power nationally. Or, to put it less cynically, it manifests itself in efforts to ensure that the philosophy of government preferred by the national party informs any decisions taken centrally about the shape of the system. Clear examples of this process in operation can be seen in the responses of Conservative and Labour governments first to the recommendations of the Redcliffe-Maud Commission and, second, to the report of the Layfield Committee on Local Government Finance.

In the case of Redcliffe-Maud, Labour welcomed the proposal for unitary authorities partly because, by recognizing the pre-eminence of urban centres as a determinant of the shape of the local government system, it would have maximized Labour's political impact at subnational level. It would have eradicated the county councils, in Labour's view the most reactionary of the existing classes of authority (Cmnd 4276, 1970). But Labour supported the proposals for a less partisan reason, too: they accorded with the party's affection for larger, better resourced and, it believed, more efficient and effective local authorities. (It should be noted in passing that democratic Socialist thinkers – the Webbs, Laski, Robson, Crosland – have been invariably vague about the effects of their interventionist social and economic policies on the fabric of local democracy. John Strachey's famous remark that 'the man in Whitehall knows best' was only the loudest note in a recurrent Labour tune.)

Similarly the Conservative response to Redcliffe-Maud was a combination of a desire to defend their centres of strength in the counties and a philosophical attachment to 'localism' as a value to be sought through the subnational system of political institutions. That attachment has been overwhelmed – perhaps temporarily – with the arrival of Thatcherism, although the occasional recalcitrance of the Association of County Councils and the Association of District Councils, and the persistence of Conservative doubts about the abolition of the GLC and the metropolitan counties suggest that local Tories do not all share the views of their national leaders. It should be noted also in this context that the 'rollback' of the state, a central tenet of Thatcherism, recognizes no distinction between elected and non-elected public authorities. The paradox of the Tory reaction to the Redcliffe-Maud report is that it led to the construction of a two-tier system that is less amenable to central control than a unitary system would have been. The plan outlined in the Labour White Paper of 1970 would have lent itself more readily to the changes in rate support grant between 1979 and 1982 than has been the case with the reorganization implemented in 1974. Also, the difficulties in the implementation of rate-capping would, conceivably, have been lessened had the number of local authorities been fewer.

When local government was reorganized in 1974, the new structure was left to operate on the basis of an archaic financing system which, itself, had been allowed to become an engine of centralization, as local government annually came to rely more heavily on exchequer contributions to their operating budget. In 1974, not because of a concern for the health and strength of local democracy but as a low-risk response to public concern about the level of local rates, Anthony Crosland appointed the Layfield Committee to examine all aspects of local government finance.

The report of that committee explicitly challenged central government, and by inference political parties, to come clean about their attitude to the central–local relationship and about their view of the place of local government in the political system. It presented two alternative futures which may be called the 'localist' and the 'centralist' options and it specified the nature of the link between decisions on how to pay for local services and the level of central control. The implication is clear: central government decisions on the nature of the grant system signify central government's view of its relationship with local authorities. Two quotations from Layfield make the point:

> A unitary grant would have an obvious place in a system based on central responsibility where the government would be responsible for deciding the main elements of local spending needs. (Layfield, 1976, p. 233)

And:

> If government assumed the main responsibility for local government expenditure, it would set totals for local spending within

> fairly narrow limits. It would have to determine how much each authority was to spend and for what purpose . . . A unitary grant based on a prescription of approved expenditure for each authority . . . would be the most appropriate form of grant in this context. (ibid., p. 293)

Neither in their formal responses to Layfield, nor in their actions in government, have the two major parties shown anything but a determination to retain the power over local authorities that is an inevitable consequence of three facts of British political life: central government's monopoly of direct taxation; the unpopularity and inflexibility of the rates; and the reliance of local government on central government grants to pay for a large part of the services they provide. Layfield's localist option would be possible only if local government were given access to a major direct tax, specifically a local income tax. But such an innovation would immediately increase the fiscal and political autonomy of local authorities and so reduce the power of the centre to control the localities. For central government (and national parties) that would increase the level of unpredictability in the achievement of goals set nationally. When given the choice, the invariable response of central government is to conserve and increase, rather than to contain and diminish, its powers over local government.

The Bureaucratic Dynamics of Reform

In addition to the centralizing tendency that arises from the political power to determine objectives and to specify the nature of institutions through which they may be reached, there is a further source of centralization: the essentially *bureaucratic* urge to control, particularly to control the standards of service and the process by which the objectives set centrally are reached. Historically, examples of this source of centralization are not hard to find. They date from the utilitarian attack on local self-government that began with Edwin Chadwick and the public health reformers in the 1840s and was pressed vigorously to a conclusion in the 1860s and the 1870s. The bureaucratic urge for control is characterized by a desire for uniformity which, in itself, is a form of centralization in that it deprives local authorities of important kinds of discretion. The achievement of acceptable standards of sanitation, sewerage and water supply did not in itself necessitate the laying of an obligation on local authorities to appoint a medical officer of health. The process of central control, however, is facilitated by the existence in local government of an official who, because he is appointed at the insistence of central government, provides a channel through which central influence, and where necessary coercion, can be exerted. The obligation to appoint certain officers was included in the Local Government Act 1972, although the necessity to have the approval of the Secretary of State for the Social Services to the appointment of a Director of Social Services, introduced as recently as 1970, was removed in 1980. It is worth noting, also, that the obligation on local authorities to appoint certain officers contributed to the process by which the relative positions in the local policy process of officers and members were completely reversed in the late nineteenth and early twentieth centuries, especially in the major urban authorities. By reducing the contribution of local politicians and increasing that of professionals whom local politicians were bound to employ, this obligation contributed to a uniformity that was conducive to centralization and control.

In the twentieth century the bureaucratic urge to control can be traced in most aspects of the relationship between central and local government. This drive may be seen in the bureaucratization of the process of boundary and status change among authorities, in the remit of the Local Government Commissions between 1958 and 1966, and in the evidence of the Ministry of Housing and Local Government and its successor the Department of the Environment to Redcliffe-Maud and Layfield respectively. It can be traced in the departmental attachment to the notion of a unitary grant to replace the locally based needs and resources elements of the rate support grant (not resisted, by the way, by either Labour or Tory ministers, probably because it is a medium for the exercise of *power* as well as *control*). Most recently, it can be traced in the enthusiasm which civil servants have shown for the grant-related expenditure assessments that are the basic components of the tight control of local authority spending and, by extension, of service levels, that is the object of, and is facilitated by, recent legislation, particularly that introducing rate-capping.

The National Voice of Local Government

The political and administrative environment in which local government has to operate, then, has been made more hostile by and since the reorganizations of the 1970s. That environment, moreover, has been conditioned by the effect of the structure of the new system on the national representation of local government. Both Redcliffe-Maud and Wheatley emphasized the need, in any new structure, for local government to be strongly represented in its dealings with the centre. Recliffe-Maud in particular (1969, para. 107) stressed how inimical to the interests of local government would be any divided system of representation. If the majority recommendations of the Royal Commission had formed the basis of the new system in England and Wales, unified representation might have been possible. For although there would have been a two-tier system in the major conurbations and a unitary system elsewhere, all the new authorities would have had in common an origin in the recognition of the essentially urban character of England. In less than abstract terms, such a structure would not have so clearly and permanently disadvantaged either of the major political parties as to move it to sabotage any proposal for unified representation. In the event, the Labour Party, both through its national organization and through its leadership of the AMC, was moved to do precisely that. Its argument was encoded in references to the need to preserve 'a distinctly urban voice' at national level. It was obvious, however, that any single association would be dominated, given the nature of the two-tier structures and the preservation of the county councils, by the Conservative Party. An association to represent the metropolitan authorities, especially if it included London, was likely to be Labour controlled more often than not. So representation was divided, or rather separated, by the creation of the Association of Metropolitan Authorities, the Association of County Councils and the Association of District Councils; this separation was further emphasized, as Gyford shows (Chapter 4), by the subsequent politicization of the major associations.

But division would not be so important if it did not also imply competition. The process of central–local relations is far from simple. It cannot be understood as a simple master and servant relationship. Nor can it be understood as a contest between two monolithic, if unequal, contestants. The division of the national representation of local government is a reflection of the structure of the local government system, which in itself perpetuates some of the divisions which it was hoped reorganization might have removed. The relevance of this division to the process of central–local relations becomes clear when discussion about local services moves from the general issue of the level of central government's financial support to local government – upon which the associations may be expected to speak with one voice – to the specific issue of how that support is to be distributed. Decisions about the balance of assistance between classes of authority have, since reorganization, been of central significance to the central–local relationship and they reveal local government to be a house divided against itself and thus at a considerable disadvantage in its dealings with the centre.

Since reorganization, a new institution, the Consultative Council on Local Government Finance, has become the primary forum for discussions between central and local government. It is composed of ministers and civil servants from the departments with local government responsibilities, and from the Treasury, and members and their advisers from the local authority associations. It is chaired by the Secretary of State for the Environment. Its establishment in 1975 was, however, announced in the Budget speech by the Chancellor of the Exchequer. This suggested that its primary purpose was not the co-operative development of local government services but the examination of their financial implications. In short, the council was an addition to the machinery of macroeconomic planning at the centre.

Despite this, it was widely welcomed in local government (Alexander, 1982b, pp. 161–3), for it gave a guaranteed point of access to replace the *ad hoc* and intermittent contacts that had existed before. Perceptions of the performance of the council were, however, profoundly affected by the expectations that its local authority members had of it. It was, as its title clearly indicates, a *consultative* institution and not a *negotiating* one. It did not proceed by agreement; its sessions were adversarial, with a local authority side and a government side; its decisions were ministerial decisions rather than agreed decisions. On the local government side, there was a tendency to talk about 'negotiating the Rate Support Grant', when all that local government could realistically hope for was the sustained exertion of its influence. That influence is, by definition, difficult to measure, but it has probably affected the process rather than the substance of central–local financial relations. All Secretaries of State for the Environment since 1973 have carried a government brief to reduce either the level of central support for local government, or total local government spending, or both. Thus, the scope for local government influence has narrowed to a concentration on ameliorating the effects of government policy.

The position of the major political parties in the process of central–local relations which condition the position of local government in the political system at large is characterized by the degree to which their activities are inhibited by the subordinate status of local government and by the existence of competing mandates. The relevance of the subordinate status of local government to this part of the discussion arises from the paramountcy given, by both major parties in government in the period since reorganization, to the management of the national economy. No matter how deeply attached the parties may have been to the notion of local government and local autonomy, a firm commitment to their defence has not been seen as compatible with national economic policy and financial management. This point is further developed, in so far as it affects the contemporary position of local government, in the final part of this chapter.

The notion of competing mandates is one that is relevant both to the local government system itself (where it may affect the relationship between the tiers) and to the relative positions of local and central government. Major local government elections take place in the middle of the life of every Parliament, often at the lowest point of a government's popularity. If local government were autonomously partisan, or if local campaigns were more often dominated by local issues, this might have little direct effect on the central–local relationship. But voters in local elections respond mainly to national cues. The political composition of the local authority associations is, of course, determined by the political complexions of their member councils. Thus, governments invariably face, at some point in their term of office, a local government delegation to the Consultative Council in which their political opponents either have a majority or have very strong representation. Either way, the political party is inhibited from taking a very constructive role in the process of central–local relations, and the definition of the place of local government is determined by the expression of the superiority of central government.

The Paramountcy of National Goals

This chapter has been concerned so far with the period between 1935 and the completion of the structural reorganizations of the early

1970s, and the nature and processes of centralization in the period up to the general election of 1979. As a prelude to the consideration of the position of local government in the 1980s, it is as well to repeat what have been argued to be the major components of the radical change in that position in the last half-century. First, the appointment of the Herbert Commission and of the Local Government Commissions in the 1950s marked the end of local government's veto power over proposals for reorganization. Second, the entire period since 1935 has been characterized by a continuation and acceleration of the process of centralization energized both bureaucratically in pursuit of control and politically in pursuit of predictable and reliable local implementation of nationally determined goals. By 1979, a third source of centralization was apparent.

This third source, which was alluded to briefly at the end of the previous section, is closely related to the desire of national government for power to ensure the implementation of its policies. It is worth considering separately, first because of its contemporary relevance, and secondly in order to demonstrate that centralization is an inherent tendency of national goal-setting, no matter the nature of the goals being set. This final source of centralization is the power, in a unitary state, of nationally determined primary goals. This power implies the capacity of the national government to subordinate means employed and effects experienced locally to ends determined nationally. A distinction is made, and this strengthens the case for considering this source of centralization separately, between specific goals, such as housebuilding (or house-selling) targets and the introduction of comprehensive education, and generalized ends, such as the conquest of inflation and control of the money supply.

This third source of centralization – the primacy of national government objectives – is the most important determinant of the changed position of local government. The argument here relates not to the constitutional superiority of national over local government but to its political expression in policies which are regarded as incompatible with the preservation and the expression of local autonomy. The relationship is subtle and to some extent dependent upon the nature of the ends sought by the centre. For where the national objective is the improvement and expansion of locally delivered public services the danger may be obscured by the fact that although national objectives may be paramount, thus reducing, or in some fields eliminating, the scope of local initiative, the product is acceptable both to local politicians and to the local population. It is when national ends are dependent upon the contraction of local services, when they are capable of achievement, in the eyes of national politicians, only by denying the distinction between central and local government, that the position of local government is profoundly altered.

In its White Paper on the future of the rating system, the government sought to justify its decision to remove the fiscal independence of local government by saying that

> Local authorities did not raise constitutional objections when central government encouraged them to spend more on particular services; objections began to be raised only after a need for retrenchment arose. (Cmd 9008, 1983, para. 1.2)

Such an argument is over-simplified, for it does not accept that local democracy is more than procedural. It does not recognize that local democracy depends upon elected local institutions retaining the capacity to do more than act as the administrative and implementing arms of the centre. The distinction between the establishment of minimum and maximum levels of local service provision and expenditure is crucial, for while the first preserves the possibility of local initiative, the second eliminates it. By operationalizing the constitutional principle of unitarism as political superiority and subordination, local democracy is attenuated into little more than the election of councillors. Local democracy is sacrificed on the altar of national policy. And that would serve as a working definition of centralization. It is a high and unacceptable price to pay for any national objective. It may be argued also that the attainment of the nationally determined goals specified by governments over the last ten years is not dependent upon greater central control of local authorities.

Even before the major changes of the past decade, local democracy was more fragile in reality than in principle. Popular participation in the local electoral process was already low. As local initiative is eliminated, participation is likely to decline further; committed councillors and able professionals, deprived of the satisfaction of serving a local population and satisfying local needs, will employ their talent and enthusiasm in other fields; national politicians will begin to question the wisdom and necessity of retaining the electoral, and hence the democratic foundation of local government. Already, one tier of elected local government is being abolished in the major conurban areas, and whatever democratic element that remained in the metropolitan counties and the GLC is to disappear entirely as services are transferred to undemocratic appointed boards. Since the overriding national objective is related to finance rather than service provision, the relationship between the centre and the localities becomes so technically complex that it leads to the progressive de-involvement of those who are democratically chosen and directly accountable to the people. That too is centralizing; that too is dangerous to local democracy.

Ivor Jennings concluded his chapter in *Municipal Progress* (p. 454) by saying that "central control" must ... involve not a meticulous investigation of the details of local administration but a general control in the national interest of divergent and contradictory local policies'. Fifty years on, that distinction is no longer recognized by national politicians. Other chapters (Jackman, Dawson) deal with the economic and financial implications of the relationship between central and local government as it has developed since the publication of *Municipal Progress*. From this chapter it should be clear that in political and administrative terms the nature of central control, where it is exercised to bring local authorities into line with the policy objectives and political prejudices of government, inexorably tends towards the 'meticulous investigation' dismissed as inappropriate by Jennings. It is worth noting, however, that Robson, in the slightly apologetic *dirigiste* spirit of the democratic socialists, was aware that in some spheres there would be in the century after 1935, 'a diminution of local autonomy, a restriction of discretion' (*Municipal Progress*, p. 461).

None of the authors of *Municipal Progress*, however, gave serious consideration to the possibility of a drastic decline in the significance, autonomy and independent discretion of local institutions. None of the authors of the present volume, had they been writing ten years ago, would have done so either. In 1975, when reorganization was completed, the local government system was at the end of three decades of almost continuous expansion of its services. Members, officers and academic commentators did not envisage the speed with which it would move first to standstill budgeting and then to contraction. Nor could they foresee how the determination of the centre to enforce the primacy of national objectives would lead to ever closer supervision, ever more meticulous investigation, ever narrowing discretion. This is the political and constitutional reality of cash limits, grant-related expenditure assessments and, most seriously of all, rate-capping. For the enforcement of national policy to contain local expenditure and to limit the level of rates has led irresistibly to closer investigation of the work of local authorities. As this is written, local government awaits the first list of authorities to be deprived of their historic right to levy a rate to finance services they consider necessary for the good government of their communities. Should they wish to challenge the decision of the Secretary of State for the Environment, they must seek a 'derogation' and if they do so, they activate his power to require from them any information which he considers necessary in deciding their application. Such information will inevitably take the form of a defence of planned expenditure on particular services. In examining the information, ministers and civil servants will push 'meticulous investigation' beyond the point where, to quote Robson again (*Municipal Progress*, p. 464), 'local government is likely to remain . . . the most effective instrument of social welfare in our national life'. Robson could not know how prescient his words, written in the shadow of rising tyranny in Europe, would seem fifty years later, when he wrote, in the penultimate sentence of the book (p. 464) that 'Dictatorship at the centre inevitably involves the arbitrary suppression of local freedom and initiative, the curtailment of local responsibility and opportunity.'

4

The Politicization of Local Government

JOHN GYFORD

> The member of a town council may be – he frequently is – a man with keen political sympathies, but these are not the dominating factors in his municipal work, and he seldom claims for them the hospitality of the floor of the council chamber, where the issues to be determined are local and practical.
>
> Lord Snell in *A Century of Municipal Progress*, p. 75

> Few questions are so intricate in character as the relation of party to committee work in local government .... Most of the large general questions which have to be solved raise issues which have a party connotation ... it is reasonable nowadays to urge that, in matters of general principle, a party atmosphere is the rule rather than the exception.
>
> Harold Laski in *A Century of Municipal Progress*, p. 83

Municipal Progress offers two different estimates of the place of politics in local government in 1935. Lord Snell plays down the role of politics while nevertheless noting the existence in some areas of 'the vote manipulator who uses purely municipal contests for party political purposes' (p. 74). For Laski however the questions raised by the issues of 'municipal trading, direct labour, the quality and character of schools, the range of public health services, [and] the attitude to slum clearance and housing' are integrally related to party politics since a council's handling of the issues depends on which party is in control (p. 85). The conflict of interpretation between Laski's emphasis on politics and Snell's downgrading of its significance was in one sense decided in Snell's favour for the politics of local government was not granted a chapter of its own in the 1935 volume and the subject appears only fleetingly throughout the book. Such hesitant or fragmentary treatment was not unusual for the time. Finer's *English Local Government* (1933) and Hasluck's *Local Government in England* (1936) were also rather reticent on the subject. So too was Maud's *Local Government in Modern England* (1932), though oddly enough Maud did talk at one point (p. 16) of local government as 'a local state', thus anticipating by some forty-five years the more recent, if rather different, use of the term (cf. Cockburn, 1977). One book published in 1935 did, however, devote a specific chapter to local party politics; the book was *How Greater London is Governed* and its author was Herbert Morrison.

In 1934 Morrison had led the Labour Party to its first victory in the London County Council elections, inaugurating three decades of Labour rule at County Hall. In that same year Labour made net gains of 666 seats in the provincial and metropolitan borough elections, giving the party control of a quarter of the county boroughs and 15 of the 29 metropolitan boroughs. (There is no single comprehensive source of local election results. I have drawn on the following sources for different periods: Cole, 1948; Maud and Finer, 1953; Cook, 1975a, 1975b; the Labour Party Annual Conference Reports for the period before 1947; the Conservative Central Office Local Government Election Results 1947–1972; and the annual Central Statistical Office publication *Social Trends* for the period since 1973.) Morrison's and Labour's victories in 1934, representing the party's best performance since 1919, can be seen to symbolize the near-completion of one particular stage in the politicization of local government, namely the replacement of the Liberals by the Labour Party as the main opposition to the Conservatives in local politics. This development, however, was only one in a number of stages in the politicization of local government in the 150 years after 1835.

Five Stages of Politicization

Looking both backwards and forwards from the vantage point of 1935, it is possible to trace within the history of local government a process of gradually increasing politicization during the century and a half after 1835. Politicization involves the growing participation of organized political groupings in such aspects of local government as

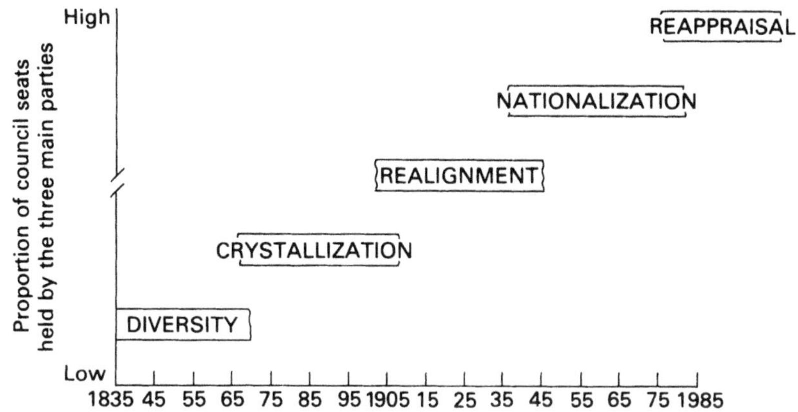

Figure 4.1: *Party politics in local government: five conjectural stages.*

local elections, the internal organization and workings of councils, the formulation and execution of council policies, and the relations between local and central government and between local government and the public. It is possible to conjecture that there were a number of stages in this process of politicization, though in doing so one runs a considerable risk of over-simplification in terms both of exaggerating the 'neatness' with which one stage gradually succeeded another and of underestimating variations in the occurrence of the several stages between urban and rural areas.

The three decades after the 1835 Act was a period of great diversity, if not confusion, with politics in local government assuming a somewhat kaleidoscopic form. A variety of actors in different towns – Whigs, Tories, Liberals, Conservatives, Radicals, Chartists, Improvers, Economizers, Independents, and a bevy of personal cliques and factions – argued over whether to expand, or to economize on, the provision of public services. This political diversity was matched by administrative diversity. There was a bewildering patchwork of functions and areas divided amongst municipal corporations, improvement commissioners, local Boards of Health and Poor Law Guardians, producing what Goschen, in a famous phrase, described as 'a chaos as regards authorities, a chaos as regards rates, and a worse chaos than all as regards areas' (quoted in *Municipal Progress*, p. 48). Thereafter, in the last third of the nineteenth century, with the larger towns leading the way, there came a stage of crystallization. In this period, despite a continued preference by some candidates for the Independent label, local politics gradually solidified into a Conservative–Liberal contest, the pace and style having been set perhaps by the impact of the 'civic gospel' of Joseph Chamberlain's Birmingham Liberalism. This was also a period in which local elections began to be seen as having national implications. Sir John Gorst, the Conservative Party's Principal Agent, began using local election results as indicators of parliamentary results in the 1870s, though Disraeli was not wholly convinced of their relevance (Hanham, 1959, pp. 388–90); and in 1894 the Marquess of Salisbury saw the LCC election as useful practice for the general election that was to follow (Young, 1975, p. 32). This was also a time when attempts were made to rationalize English local government. The early 1870s in particular were 'something of a watershed' (Fraser, 1979, p. 152). A Local Government Board was set up in 1871, which combined the functions of the Public Health branch of the Privy Council, the Local Government section of the Home Office and the Poor Law Board, and a uniform pattern of public health administration emerged from the 1872 Public Health Act.

After this era of political crystallization, accompanied by administrative rationalization, the creation of the Labour Party early in the new century marked the beginning of a period of realignment. The nature of local politics became transformed as Labour replaced the Liberal Party, gradually at first but with gathering pace during the 1920s and 1930s. This stage of realignment in local politics was followed, in turn, after the Second World War, by one of nationalization, in which local politics became harnessed, partly in aspiration and partly in actuality, to the national fortunes of the major parties, with local government itself becoming an important arm of the postwar welfare state. Finally, in the wake of local government reorganization in the early 1970s, though by no means simply as a result of it, there developed what might be described as a stage of reappraisal in local politics, with the emergence of major debates both between and within the political parties about the role of local government in society.

Each of these different stages was basically concerned with the form, rather than the content, of political conflict at the time. Yet form and content were interrelated, and both in turn were influenced by developments in the wider political and social systems, as can be seen more clearly in the history of events between 1935 and 1985.

Realignment

The *Municipal Progress* was published during the middle or third stage, that of realignment. Labour councillors had already made modest inroads into local government in the years before 1914 and, indeed, had taken control of West Ham briefly in 1898 and of Woolwich in 1903. Occasionally, Labour candidates stood in alliance with the Liberals, as in the case of London's Progressive Party, but many of Labour's pre-1914 gains were won by candidates sponsored by the Independent Labour Party (ILP), the Social Democratic Federation (SDF) or local Trades Councils, who stood in their own right; and Liberals in towns such as Sheffield, Bristol, Coventry and Leicester were soon seeing Labour as a threat to their councillors. Cole (1948) estimated that there must have been about 420 Labour councillors elected under a variety of labels in the provincial and metropolitan boroughs by 1914 with a much larger but unspecified number on district and parish councils and Boards of Guardians. Local elections were suspended during the war, but the near-doubling of the local government franchise in 1918 and the Labour Party's adoption of a constituency-based structure in the same year paved the way for a further advance at the expense of the Liberals in the interwar period. So too, of course, did the lingering divisions in the Liberal Party over the wartime Lloyd George coalition.

Max Beloff (1975, p. 8) has described the history of interwar local politics as being one in which 'Liberal councillors succumbed to the Tory embrace while their old following among the electorate largely drifted into the Labour camp'. More specifically, by drawing on Cook's analyses of interwar electoral politics, it is possible to identify six major trends in this period of realignment (Cook, 1975a, 1975b). They are: a slow but persistent fall in the number of Liberal candidates in municipal elections (from 18·6 per cent of the total in 1921 to 5·6 per cent in 1938); an uninterrupted net loss of Liberal seats each year; an increasing number of Labour candidates; a series of net Labour gains in the municipal elections in all the interwar years other than 1922, 1930, 1931 and 1936; the emergence of anti-Socialist alliances, uniting Liberals and Conservatives and sometimes non-party Ratepayers, to meet the Labour threat and, in the middle and late 1930s, a rise in the number of Independent candidates, possibly the result of erstwhile Liberals rejecting both their own party label and the Conservative embrace of the anti-Socialist alliance. The municipal elections of 1945 set the seal on this period of realignment. They were the first to be held under universal adult suffrage for the local government vote and they saw Labour offering 46·2 per cent of all candidates and securing 62·8 per cent of all councillors. For the Liberals, the respective figures were 8·0 per cent and 4·2 per cent.

The causes of Labour's replacement of the Liberals were bound up with events of more than local significance. Nevertheless the Liberal predicament was not assisted by certain shortcomings of their own at the local level. The party had little in the way of a distinctive municipal policy; its old role as standard-bearer of chapel politics in education was rendered irrelevant as Nonconformity had dwindled as a major social force in many towns; an enthusiasm for 'economy' as distinct from 'extravagance' made it appear to echo the Conservatives and the Ratepayers; the party either lacked or rejected younger, working-class candidates, offering instead older candidates from the commercial and shopkeeping classes; and on key topics such as housing and public assistance it had little to offer by way of policy which might tempt the potential Labour voter. Increasingly local politics was coming to be divided, like parliamentary politics, along class lines (Wald, 1977), thus foreshadowing the two-party politics of the post-1945 era.

Yet in this interwar period, municipal and national politics were not wholly assimilated into one another. For example Maud (1932, p. 81) observed that the correspondence betweeen the results of local and parliamentary elections at this time was 'erratic'. He later concluded that 'the only election of that [interwar] period in which such a correspondence quite clearly took place was in 1931, where the overwhelming defeat of the Labour party in the general election was followed by a similar reaction in the municipal boroughs' (Maud and Finer, 1953, p. 85). A similar assessment has been made by Cook, who noticed that Liberal 'revivals' in parliamentary contests in 1923 and 1927–8 were not matched in local elections. Additionally, he found 'a fundamental divergence in British politics in the period after 1931 between Labour's performance in general elections and its performance in municipal voting', with Labour's rapid recovery at municipal level failing to express itself to the same degree in parliamentary by-elections or at the 1935 general election (Cook, 1975b, p. 172).

It is thus possible to see local politics in these years as having, in some degree, a life of its own and, indeed, one social historian has suggested that 'the period from the late nineteenth century to the Second World War might well be called the heyday of local government' (Stevenson, 1984, p. 309). Local government took charge of education with the abolition of the School Boards in 1902 and of Poor Law duties, including some hospitals, with the abolition of the Boards of Guardians in 1929. Moreover, despite a vigorous controversy over municipal trading in the early years of the century, local authorities still retained the public utility functions of gas, water, electricity and transport provision pioneered by their Victorian predecessors. One consequence of this range of duties was that local authorities were major employers not only of white-collar but also of blue-collar workers. They were thereby contributing, both through services and through jobs, to the welfare of a substantial cross-section of the community and were less open to the charges, made in the later era, of being all Chiefs and no Indians.

Those manual workers who were employed in the local authority utilities were, however, precluded from serving on the authority as councillors, a particularly significant barrier in the single-tier county boroughs. This prohibition may have delayed somewhat the advance of manual workers into council membership until after the utilities were removed from local government in the period from 1945. In any event 'the inter-war years have been called "the golden age of the shopocracy" in local government' (Stevenson, 1984, p. 314), with shopkeepers and small businessmen succeeding to the role previously played in local councils by industrialists or landowners (Lee, 1963; G. W. Jones, 1969). The 'territorial abdication' (Cannadine, 1982, p. 5) of the landowners from local politics became especially marked as the great estates were broken up and sold off in the years following the First World War in what has been described as the greatest transfer of land since the dissolution of the monasteries (Thompson, 1963). Although certain landed interests, notably farmers, were still to be found in rural local government well into the post-1945 era, by 1939 the process was virtually completed whereby those 'social leaders' who had once graced local government with their own high social status had been replaced by somewhat more humble 'public persons', who acquired status through, rather than bestowed it upon, their local council (Lee, 1963). Humble though they might have been, however, these 'public persons' were often the social equals of the council officers with whom they dealt, for at this time local government officials were still a largely non-graduate, non-professionalized group, a fact which enabled members and officers to meet on something like equal terms.

Nationalization

It is difficult to know in detail exactly what the overall political complexion of local government was in 1939. By then, Labour claimed to hold over one-third of all county borough seats (including Aldermen) and about one-sixth of non-county borough seats. Labour controlled 4 county councils (Durham, Glamorgan, London and Monmouth), 18 out of 83 county boroughs, 24 out of 273 non-county boroughs, 17 out of 29 metropolitan boroughs and over 100 urban and rural district councils. When local elections resumed in 1945, Labour advanced strongly from its 1939 baseline. The elections gave Labour control of 10 county councils, 47 county boroughs, 73 non-county boroughs, 23 metropolitan boroughs and 150 district councils. Moreover, whatever the pattern may have been before the war, this local advance paralleled a national parliamentary advance. Indeed, the two sets of elections were linked specifically in Labour's appeals to the voters at the local elections, with a Transport House broadsheet reminding them that 'the Labour Government at Westminster depends upon your council' and with the party's National Agent referring to local authorities as bodies who were 'going to administer the socialist legislation the Government will put through'.

The years after 1945 saw a major escalation of party politics in local government and a concomitant decline in the number of Independent councillors. In the years between 1947 and 1952 the Independents suffered an unbroken series of annual net losses totalling some 1,500 seats in the county and municipal elections. In the following decade they lost a further 340 seats. The spread of party politics was not uniform but occurred largely in the urban areas. By the time of the last local elections before reorganization, in 1972, the Independents had been reduced to holding only one seat in nine in the county and non-county boroughs. Their fate here differed from that in the county councils, where they were able to hold approximately 30 per cent of all seats throughout the eight triennial elections from 1949 to 1970. Thus, the overall position as it developed after 1945 was very much that discovered by the Maud Committee inquiry in 1964–5: three out of four of the boroughs and urban districts were run on a party basis compared with only one-third of the counties and fewer than a quarter of the rural districts (Maud, 1967, pp. 108–9).

The expansion of party politics after 1945 was accompanied by a certain assimilation of local political battles into the national contest between the two major parties, a process later christened the 'nationalization of local politics' (R. W. Johnson, 1972; Young, 1975; Schofield, 1977). An important component of this process was a growing and more systematic involvement with local government on the part of the national leaderships at party headquarters and Westminster. The Conservative Party set up a National Advisory Committee on Local Government in 1944, created its own Central Office Local Government Department in 1946, began to convene annual local government conferences for Conservative councillors in 1947 and launched a monthly journal, *The Councillor*, in 1948. On the Labour side, despite an earlier initiative in 1936 in establishing a Local Government Department at Transport House, the inauguration of a *Newsletter for Labour Groups*, the setting-up of an NEC local government subcommittee and the initiation of an annual local government conference did not occur until 1954, 1955 and 1956 respectively.

Meanwhile, however, both parties had seen ways in which local politics might be played in aid of their national struggle. After Labour's campaign attempted to link the two levels of politics in 1945, the Conservatives were not long in replying in kind. The 1947 local elections saw substantial Conservative gains, which Central Office felt laid the foundations of eventual recovery at parliamentary level after the disaster of 1945. The 1947 gains were due not to a decline in Labour support (which actually increased in terms of votes cast) but to an even larger increase in Conservative votes. The voters, moreover, were responding to a Conservative campaign that had been waged not only on a 'Town Hall versus Whitehall' theme of defending local autonomy but also on national issues, such as food shortages, the abolition of the basic petrol ration and the general 'austerity' which accompanied the export drive.

The 1950s and 1960s saw repeated attempts to bring local and national politicians of each party to march in step. The calls for greater co-ordination did not come solely from the top; from the bottom, from the councillors, came intermittent pleas for ministers of each party to bear in mind the electoral and other problems of their respective local colleagues, for example, in the timing of controversial policy initiatives. The 'marriage' of local and national politics was perhaps best symbolized by the decision of Conservative Central Office to exploit the party's massive gains in the 1967 local elections by taking control of the Association of Municipal Corporations, which had previously been run on largely non-political lines. In a concerted campaign, the more numerous Conservative-controlled boroughs represented on the association succeeded in capturing all but two of its thirteen committee chairmanships. The eventual outcome of this coup was the conversion of the local authority associations, before and after the 1974 reorganization, into political bodies to a previously unknown degree, with Labour retaliating in kind when electoral circumstances allowed.

A noticeable feature of the three decades after 1945 was the extent to which local voting patterns seemed to be largely a product of national trends rather than local issues, a phenomenon which had been observed in the nineteenth century but which had declined during the period of realignment. Its re-emergence meant the appearance of marked peaks and troughs in the local electoral cycle, as councillors across the country reaped the reward, or paid the price, of their national party's standing in the voters' esteem. The most marked of these, after Labour's triumphs in 1945, was the Conservative victory of 1968 when the party gained 1,630 seats in the provincial boroughs and urban districts, with Labour losing 1,602. An important consequence of these developments in 1968 was the sweeping from office of many long-entrenched Labour councillors and the undermining of the traditional right-wing Labour leadership on many authorities, thereby opening the way for a new generation of local Labour politicians in the years which followed. The explanation for such electoral events lay in a tendency noted by Butler and Stokes (1975, p. 40): they found that voters in council elections 'voted to an over-whelming degree in line with their expressed [national] party self-image . . . well over 90 per cent of our respondents stayed with their generalised ties to the national parties'. In such circumstances local elections were not so much a test of opinion on local issues but more an opportunity to exercise the party machine, a fact which led party organizers to favour annual local elections in particular.

Butler and Stokes' references to the voters' 'generalised ties to the national parties' reminds us that they were studying political behaviour within a stable two-party system wherein each party enjoyed a large block of 'captive' potential voters, whose loyalties were chiefly influenced by factors of family and class, leaving electoral outcomes to be heavily affected by differential abstention and a minority of floating voters. Moreover, the majority of the politicians who operated within this system, whether at the local or national levels, inhabited the same 'assumptive world' (Young and Mills, 1980). They took for granted the permanency of the welfare state and the mixed economy so that their partisan disagreements covered a rather narrow range of opinions. Such an era of stable consensus politics, sustained by a level of economic growth which seemed to provide resources for many otherwise competing demands, was fertile ground for the nationalization of local politics.

Within local government itself political debate tended to revolve around what policies to adopt within the framework of the welfare state, for example around issues such as the 11-plus examination or rent levels rather than around more basic topics, such as the merits of state education or of council housing. There were few serious ideological stresses and strains within either of the parties and little disposition to raise fundamental arguments about the respective roles of local and central government.

In such circumstances, the nationalization of local politics, with its overtones of 'homogenization' (Bristow, 1978) and 'a gradual ironing out of autonomous local characteristics' (R.W. Johnson, 1972, p. 53), can be seen as the particular expression in local government of that bipartisan collectivism which Samuel Beer (1965) identified as being the then distinctive feature of British politics. As such, it is not to be equated with the demotion of local authorities into mere tools of central government, with all power centralized in Whitehall. As recent research has suggested, the central–local relationship has been more complex; councils have had some degree of room for manoeuvre (Rhodes, 1981). The nationalization of local politics rested less on organizational or administrative mechanisms within parties or public authorities and more on the shared assumptions and values of local and national politicians. It represented one aspect of the mid-twentieth century *rapprochement* between the feudal paternalism of Tory Democracy and the Fabian centralism of Social Democracy.

The impact of party on local government in this period became a matter of considerable academic interest. Party was becoming an increasingly pervasive phenomenon, but were there any discernible consequences for local authorities? Observers concluded that parties did fulfil certain functions. They acted as channels of recruitment for council candidates and, arguably, enlarged the pool of potential councillors; they represented different social groups and interests in the community; and they could, through their party groups, give some coherence to the internal workings of councils and their committees (Lee, 1963; Bulpitt, 1967; Wiseman, 1967; G.W. Jones, 1969; Hampton, 1970; Dearlove, 1973). Moreover their expenditure patterns did show some differences, with Labour authorities spending more readily on ameliorative and redistributive services and Conservative authorities doing likewise for nondistributive services such as police and highways (for example, Alt, 1971; Boaden, 1971; Sharpe and Newton, 1984). Although they were often keen to support the expansion of existing services, when it came to initiating new policies local party groups often seemed either unable or unwilling to go very far. In what was perhaps one of the major instances where local authorities undertook new ventures – high-rise housing and town centre redevelopment – both Labour and Conservative councils alike, in getting caught up in the spirit of large-scale planning and modernization, were responding as much, if not more, to pressures from the professional, commercial and industrial sectors and from central government than to pressures from within their own ranks or from their own local party branches (Dunleavy, 1981).

The politics of local government in the three decades after 1945 was thus in most cases closer to what Hill (1972) described as 'administrative politics'. Administrative politics is found where the administrators dominate the decision-making process, yet acknowledge the claims of democratic politics by bargaining, negotiating or consulting with people outside their own organization, and, when necessary, securing legitimacy for their own actions through the use of formal democratic mechanisms. In such circumstances politicians play two roles. They are seen by the administrators as representatives of public opinion in general or of particular interest groups. As such they must be consulted and policies must either be sold to them or constructed so as to avoid giving political offence. Additionally individual politicians may become, in effect, co-opted into the ranks of the administrators, often becoming 'socialized' into the professional ethos of the education officers, or the planners, or the housing managers. As the local government service became, after 1945, increasingly the preserve of graduates and professionals the possibility of less well-qualified part-time and unpaid councillors becoming co-opted or dominated by their officers grew increasingly real. In such circumstances, politicians might certainly scrutinize their officers' policies, and might accept, modify or occasionally reject them; but rarely would they initiate any of their own. This statement is not to suggest that initiatives never occurred, but rather that they rarely originated with the politicians (Green, 1981). Hill speculates, for example, that securely entrenched political control in one-party authorities might merely provide 'the stability necessary for administrators to take bold policy initiatives under the protection of a stagnant political situation' (Hill, 1972, p. 231).

Hill contrasts administrative politics with ideological politics in which 'unified political parties provide specific programmes' (ibid., p. 211). The contesting of elections on a party basis, and the operation of a party group system within the council, does not, of itself, signify the existence of ideological politics since the majority party may content itself merely with legitimizing the policies prepared by the officers rather than with formulating its own. It is the promotion by the party of a set of distinctive policies based on its own ideology which is the particular feature of ideological politics. This latter practice, however, was to become increasingly common in local politics in and after the mid-1970s, with Labour leading the way in the production of often very detailed policy manifestos for the larger authorities (Fudge, 1981; Game and Skelcher, 1983).

Reappraisal

In the years since the 1974 local government reorganization, some changes in local government suggest that another distinctive stage in its political history may be unfolding. One clear trend has been a further escalation in the spread of party politics. By 1983 85 per cent of all councillors in England and Wales were representatives of the

Conservative, Labour or Alliance parties. Only in the non-metropolitan districts did the three parties account for less than that proportion, and even there the figure was 80 per cent. In the metropolitan counties and districts and the Greater London area, the proportion was 98·2 per cent. The continuous advance of party since the first elections to the reorganized authorities in 1973 has been sustained as an act of deliberate policy by the parties themselves. It has, however, been taking place under very different political circumstances from those which attended the previous escalation of party conflict after 1945.

Local party politics after 1945 took place inside the confines of a stable two-party system operating within a widespread welfare state consensus, providing a localized version of 'Butskellism'. The most recent phase of local politics has emerged at the same time as a number of developments very different from those of the consensus years. These include: a greater electoral volatility accompanying class dealignment, with public/private sector location, union membership, and, in some localities, ethnicity becoming important factors in voting behaviour; the emergence of a strong third-party challenge to the two major parties; ideological polarization within the Conservative and Labour parties; an emerging debate about the future of the welfare state; and persisting economic difficulties.

These last two developments were of considerable consequence for local authorities, as they created pressures for a major reappraisal of the role of local government. The desire of successive governments to overcome the nation's economic difficulties by curbing public spending made local government the target for measures designed to control all local authority spending, including even that financed from the rates. In addition, local government found itself embroiled not only in arguments about public expenditure but also in a wider debate about the appropriate role of the state. After a generation of widespread acceptance of the welfare state there emerged in the 1970s a growing tendency to question its virtues from a wide variety of points of view. Local government became one of the objects of this questioning.

Local government has thus found itself in a position of some political insecurity since 1974, the object of often sceptical, and sometimes unfriendly, comment. Growing electoral volatility has made for more rapid changes in party control of councils, one consequence of which has been what Hampton (1980, p. 5) describes as 'a decline in the mellowing effects of several years' socialisation of newly elected councillors'. Cut-backs in services whose provision had come to be taken for granted in an earlier period aroused concern amongst those affected, as did the issue of rate levels, and some evidence has emerged suggesting that local elections might now be rather more affected by local factors than in the immediate past (Gyford, 1984, pp. 117–19). Meanwhile, as the parties themselves, even allowing for the switchbacks of electoral fortune, became more deeply entrenched in local government, the two major parties experienced a degree of ideological polarization, with economic individualism emerging as a dominant philosophy on the Conservative right and Marxism on the Labour left. Thus, even local councils remaining nominally under the control of the same party experienced pronounced switches of policy as one party faction gained the upper hand over another. In other cases, the growth in the number of Liberal councillors, from 897 in 1977 to 2,010 in 1984, and the emergence of the Social Democratic Party (SDP), with 302 councillors in 1984, have led to the creation of 'hung' councils with no one party in overall control. Hung councils had in any case been rendered more likely after 1974 by the abolition of aldermen whose six-year term of office had sometimes enabled parties to cling to power even when defeated at the polls. The various changes of control and of direction flowing from this political instability have in turn provided further fuel for arguments about what those in charge of local councils think they are doing, as the 'mellowness' of earlier years has given way to the more vigorous cut and thrust of inter-party and inter-factional contests.

The embattled image of local government has been further sharpened by the way in which the conflicts between central and local government, inherent in the former's attempts to control the latter's expenditure, have been overlain by both inter- and intra-party disputes. Thus, in the debates over the Conservative government's policy on local authority spending after 1979 the government's stand was under constant sniping from some Conservative local politicians (and, indeed, back-bench Conservative MPs) defending the prerogatives of local government; at the same time the Labour Opposition spokesmen came under fire from some Labour councillors for failing to secure a more concerted opposition to the spending cuts amongst Labour-controlled authorities.

These internal party debates have reflected the fact that none of the parties possesses any machinery by which local councillors can be forced to comply with the wishes either of the party at the centre, be it at headquarters or in the House of Commons, or of the party at large convened in annual conference. In the party constitutions there are indeed provisions for making councillors accountable in varying degrees to local party units but none for making them accountable to any wider or superior constituency within the parties. In so far as it expressed itself in harmony between a party's councillors and parliamentarians, the earlier nationalization of local politics was a product of shared values and ideas rather than of formal constitutional mechanisms within the party. The tendency of the parties to appear to work in a unitary fashion in the 1950s and 1960s was a reflection rather than a cause of widespread agreement on the issues of the time. The rebirth of ideology in the 1970s produced strains within both the Conservative and Labour parties. These tensions became particularly evident as local and national politicians of both parties tried to cope with the Conservative legislation on local government after 1979. The divisions which this legislation produced within the two major parties revealed them each to be, in the words of Richard Rose (1980, p. 254), 'a complex set of plural institutions' within which contending views of local government came to be expressed, with radicals of both right and left developing their own distinctive approaches.

These two radical approaches evince certain parallels. Both view local government in the context of its relationship to the economy, the radical right picturing it as an impediment to the fulfilment of the promise of the free market, the radical left seeing it as caught up in the social conflicts generated by capitalism and, thus, as a possible arena for opposing the capitalist system. Both express hostility towards the managerial power of the professional local government officer. From the radical right, Michael Forsyth (1981) deplores 'unnecessary layers of bureaucracy' and 'impersonal, sneering professionals', while from the radical left, Ken Livingstone (1982, pp. 34–5) complains that 'what is easy for the bureaucracy takes precedence over everything else' and that 'we are drifting into a dreadful bureaucratic paternalism'. Both express a concern for greater participation by the public, though one seeks it through individual consumer control in a free market and the other through an expansion of collective decision-making at the grass roots of politics. Both also desire to see local government in some way 'opened up', either to the impact of market forces or to the influence of popular political pressure.

None the less, these two lines of thought are heading in different directions and lead to highly contrasting policies when the Conservative right or the Labour left take control of local authorities. By early 1983 over two dozen Conservative councils had contracted out to private enterprise such services as adoption, community care of the elderly, highway maintenance, legal services, rent collection, street cleaning, pest control and recreation management. By mid-1984, one council – Merton – had privatized street cleaning, refuse collection, school meals, school cleaning and meals for the elderly and was contemplating the transformation of its finance department into a private company. Left-wing councils meanwhile were concentrating on the development of a 'local socialism', embracing such features as decentralization to ward or neighbourhood offices or committees, economic intervention through local enterprise boards, financial support for radical community and minority groups, nuclear-free zones and the mobilization of popular support for local services faced with government-inspired cut-backs (Boddy and Fudge, 1984). Such policies, moreover, were not the creation of local authority officers, as under 'the administrative politics' model. Officers might advise, but these policies represented initiatives stemming from political sources. In addition, the policies were increasingly likely to have been put before the local electorate in the form of detailed manifestos running sometimes into scores of pages and thousands of words. On the Labour side at least, the party outside the council would be likely to see itself as 'guardians of the manifesto', keeping up the pressure on its councillors to ensure that party policy was implemented. In such developments we see the incursion of Hill's 'ideological politics' in which parties 'provide specific programmes' (Hill, 1972, p. 211).

Hill (p. 216) has observed that ideological politics 'puts great strains upon administrators'. It seems likely that the strain may be all the greater if the administrators have been brought up within the tradition of administrative politics. Certainly one development of the past few years has been a questioning in some quarters of the traditional model and role of the non-political local government officer. A number of town planners formed a Radical Institute Group within the Royal Town Planning Institute to promote a more egalitarian approach to managing urban change. Some planners have decided to serve as officers in one authority and as councillors in another. This practice has been by no means confined to that profession alone in recent years, especially since the replacement of county boroughs by two-tier local government in 1974 made it a practical proposition in the major provincial conurbations. Social work too has become increasingly conscious of a political dimension to its work. A measure of this can be seen in a report by the National Institute for Social Work in 1982 which accepted that social workers saw 'links between social policies and the problems clients bring' and that it was not their role 'simply to accept things as they are'. 'The "political" element of their work is therefore an integral part of the whole . . . . We would urge elected members . . . to ensure that social workers have channels open to them . . . to bring their assessments of the effects of policies forward.'

Other departures from traditional practice have included the appointment of staff who are known to be politically sympathetic to the politics of the majority party and the breaking down of the traditional division of labour between members and officers by involving both of them in research and report-writing. The campaign for NALGO affiliation to the Labour Party, so far unsuccessful, may be seen as a further attempt to break with past practice. A leading pro-affiliation campaigner amongst NALGO members, Derek Hatton, an officer of Knowsley District Council, later emerged as Deputy Leader of the majority Labour group in Liverpool, thereby integrating within his own person the four roles of officer, trade-unionist, party activist and councillor. He was also one of the growing number of assertive full-time councillors who were as much at home in the town or county hall as the officers thanks to the generosity of their employers and/or to the payment of councillors' allowances.

The case of Liverpool also highlights another departure from previous practice in respect of local authority trade-unionism, namely the attempt to involve the unions more closely in the decision-making processes of the council. Thus, in June 1983 the council agreed that its Personnel Committee should include twelve members of the Joint Shop Stewards Executive Committee as non-voting advisory representatives and that the unions should be represented on all interviewing panels for vacancies and promotions. The Liverpool arrangement was not wholly innovatory. A survey for the Society of Local Authority Chief Executives in 1976 found that

Hereford and Worcester, Basildon and Slough made provision for non-voting employee representatives on some committees in addition to the more widespread use of Joint Consultative Committees and a number of other authorities have since followed suit. In addition to participating in such formal arrangements, the trade unions have had an impact on local government through their growing readiness to take industrial action, with the 1970s seeing the first official strikes called by the local government unions on a nation-wide basis. Initially, this increased militancy amongst local government workers reflected reaction to what was seen as central government interference in local authority pay bargaining and in the established patterns of industrial relations and trade union rights; subsequntly it also acquired another dimension, that of defending jobs threatened by cut-backs in local government spending (Walsh, 1982).

The role of trade-unionists in local politics is not confined to the stage provided by local government itself. The presence of members and officials of NALGO, NUPE, NATFHE and the NUT on local Labour Party committees whether as union delegates (in the case of NUPE) or as party branch members has grown with the expansion in public sector employment and unionization and in white-collar activism within the party. Some observers have noted the possible emergence of a public sector or public service 'class' of Labour councillors, often quite as well qualified academically and professionally as the council officers who serve them, and drawn disproportionately from the ranks of education, local government, trade unions, public corporations and publicly funded voluntary and community groups (for example, Lipsey, 1982; Walker, 1983).

The place of voluntary and community groups in local politics has aroused increasing interest in the wake of major programmes of funding by some Labour councils. For example, in 1983–4, the GLC's Grants Subcommittee authorized 184 grants costing nearly £3 million in addition to grants made by the major committees of the GLC; the Women's Committee granted £3·14 million to 216 groups in 1983–4 and the Housing Committee £4 million to 280 groups in the same year (GLC, 1984).

It should not, however, be supposed that public sector professions, trade unions and voluntary groups were alone in being drawn in varying ways into local politics. The Conservative government was simultaneously introducing legislation making it obligatory for a local authority to consult industrial and commercial ratepayers about its expenditure plans before it could set its budget. An even more direct commercial involvement could be seen amongst those private firms which received the contracts to carry out local authority services subject to privatization by Conservative councils.

The common theme these developments represent, over and above the growth of ideological politics, is a move away from a form of what C. P. Snow called 'closed politics' towards a more 'open politics', in which the authority's political links with elements of the wider economic and social structure became important features of its decision-making. The openness is not however total, since the political complexion of the council may determine which elements in the community are afforded more open access to the authority. Changes in the style of local political leadership have also taken place, with strong and sometimes rather secretive leadership being replaced by a more open and democratic system of decision-making within party groups in which leaders often become brokers between different factions or interests (G. W. Jones and Norton, 1979; Elcock, 1981). It is therefore not surprising that a major body of academic research after the mid-1970s focused not simply on the local authority itself and its inner workings but also on the place of the authority, and indeed of local government itself, in its wider setting. Studies of urban social theory, of local political economy and of central–local government relations were all part of an attempt to deal with, amongst other things, the external politics of local authorities embracing their relationships with social classes, local party organizations, economic interests, voluntary bodies, professional bodies, party headquarters and Whitehall departments (Dunleavy, 1979, 1981; G. W. Jones, 1980; Newton, 1980; Saunders, 1980, 1981; Green, 1981; Rhodes, 1981; Cousins, 1982; Gyford and James, 1983; Young, 1983; Young and Mason, 1983).

In moving from closed politics towards open politics, local government might also find itself undergoing a further change in addition to the move from administrative to ideological politics. If service delivery through privatization and decentralization to wards or neighbourhoods were to gather pace and if local authorities were increasingly to become sources of contracts, franchises and grant funding, then the tasks of co-ordination and resource allocation and of dealing with clients and agents might begin to look remarkably like the bargaining politics which Hill (1972,, pp. 217–26) noted as particularly characteristic of US local government; they might also take on overtones of that other traditional feature of US local government – patronage politics. Such developments might be anticipated particularly in those urban areas where ethnic, religious and other minorities exist in sufficient numbers to make them worth wooing by local politicians impressed by their potential voting strength.

At present it is too early to be certain how events will develop during the current stage of reappraisal. Ideological politics, open politics, bargaining politics and patronage politics in some combination may come to supplement if not wholly replace administrative politics. The political debate on the role of local government during this period of reappraisal seems to be just as vigorous as that which attended the work of the pre-1835 Municipal Corporations Commission. That however, is just as it should be, for the current debate is essentially about the very issue which the Commission's secretary Joseph Parkes identified as the fundamental question: it is about 'what our civic institutions are and what they should be' (quoted in Fraser, 1979, p. 152).

Acknowledgement

I am grateful to George Jones for his helpful comments and suggestions on a draft version of this chapter.

5

The Functioning and Management of Local Authorities

J. D. STEWART

> But all things considered, the committee system has proved itself amply in the working. The technique of the committee will certainly be found to be the pivot which makes possible the democratic operation of local government.
>
> Harold Laski in *A Century of Municipal Progress*, pp. 107–8

> Two essential principles recognized by all local authorities together make for the soundness of local administration; the first is that positions in local government are regarded as 'careers', and appointments are considered to be for a lifetime. The second is the absence from such appointments of influences associated with political interests. The local government officer who carries out his responsibilities to the best of his abilities and recognizes that he is a servant of the local authority is never influenced or affected by changes in the political colour of the governing body.
>
> L. Hill in *A Century of Municipal Progress*, p. 113

These few passing references apart, the authors of *A Century of Municipal Progress* paid virtually no attention to the administrative processes of the local authority. It is as if they were seen as non-problematic. Administration was taken for granted and municipal progress was concerned with the passage of legislation and the decisions of councils. That perspective was not universally shared by other writers at the time. *Municipal Progress* was published at about the same time as Harry Page's *Co-ordination and Planning in the Local Authority* (Page, 1936), which focuses on the administrative process. Furthermore the pages of *Public Administration* in the 1930s are full of articles highlighting problems of administration in local government.

In fact the process of administration can be seen to present no problems only if it can be assumed that the administrative process automatically ensures the effective carrying out of the statutory obligations of the authority in accordance with the decisions of the local council. That, however, cannot be assumed. The neutrality of the administration cannot be taken for granted, and its efficiency cannot be regarded as unproblematic. Over and above questions of efficiency the characteristics of the administration can have an influence in their own right both on the decisions of the council and on the way the decisions are carried out. The nature of the administration can shape the activities of the council as surely as political choice. There can be a tension between the nature of a local authority as a political institution constituted for local choice and the administrative structure and processes moulded by the requirement of existing services.

Consequently, the problem of administration has been a subject of continuing debate and discussion throughout the last fifty years. As the years have passed the language of debate has changed. The problem of administration has become the problem of management. Yet beneath the changing language (and in this chapter both words are used as a sign of the change) there have been continuities. This chapter will therefore discuss the continuing tradition of local authority management practice, will examine the main movements for management change, and will assess their impact and achievements.

The Continuing Tradition of Managerial Professionalism

Local government management structures and processes are based both on bureaucracy and on professionalism. Management practice in local government may therefore be characterized as managerial professionalism. Managerial professionalism draws upon both bureaucratic authority and professionalism and to a large extent professionalism and bureaucracy reinforce each other.

There are three dominant organizing principles in the continuing tradition of local authority management: the principle of functionalism; the principle of uniformity; and the principle of hierarchy. In the first two principles bureaucracy and professionalism reinforce each other, while in the principle of hierarchy professionalism and bureaucracy modify each other. It is through the operation of these principles, therefore, that the nature of managerial professionalism may be discerned.

The principle of *functionalism* is the division of the local authority's organization into its major functional responsibilities, such as education, social services, highways and housing. The functional divisions are written into the structure of departments for the administration of the authority's services. The same functional divisions have normally been written into the structure of committees, as the setting for political decision-making. Chairmen of service committees and chief officers of departments have forged a powerful alliance in support of the functional divides of local government.

In each department there is a dominant profession. This does not mean that all staff in that department belong to that profession, for many may belong to other professions or to none. It means, however, that the dominant positions in the department are held by a particular profession and the culture of the department reflects the viewpoint of the profession.

Any large-scale organization divides itself up to do its business. Functionalism is the way the local authority does so. That functionalism is written deeply into the management of local authorities because department, committee and profession reinforce it. Over the fifty years the functional divisions have changed with new legislation building up some services and removing others. As a result the professions have changed in status. In the lists of principal officials of the ten main cities of fifty years ago, for example, one would not find a Planner or Director of Social Services (G. M. Harris, 1939). Those functions have been established since the war and professions have grown in status to match them. One would, however, find the General Manager of the Gas Department and the Medical Officer of Health (representing functions since removed from local government). Nevertheless, whatever changes have taken place in the range of functions operated by local authorities the principle of functionalism has remained a dominant organizing principle.

The principle of *uniformity* requires that so far as possible the services of the authority are provided to a common standard and on a common pattern throughout the area of the authority. Bureaucracy flourishes by the application of common rules throughout the area of the authority. Yet both professional and political perspectives can also support the principle of uniformity. Professionalism has its own standards and standards can enforce uniformity. Political priorities can also enforce uniformities as the council seeks to establish its policies throughout its area.

Since the 1930s, with local government reorganization, local authorities have grown in scale and in the diversity of areas covered. The 1974 reorganization in England and Wales brought together town and country in the new authorities. The new local authorities often brought together the areas of several former authorities, each of which had its own policy. The response of the new authorities was to establish, as a matter of priority, uniform policies throughout the authority. The principle of uniformity was so deep in the management processes of local government that the very identity of the new authority seemed to depend upon it.

The principle of *hierarchy* is the division of the local authority's organization into tiers of responsibility. Within the functional divisions represented by the departments of the authority staff are arranged in a hierarchy of accountability, running from field-worker to chief officer and through the chief officer to committee and council. Hierarchy ensures formal control within the bureaucracy of local government.

Spans of control in local government are often limited. As the scale of local government has grown over the last fifty years with a growth in staff from under 1 million to over 2.5 million, the response has been to increase the number of tiers in the hierarchy. With more staff to control, more tiers were added to the hierarchy – as they were in those authorities whose scale increased at local government reorganization. It was the inevitable response given the acceptance of narrow spans as necessary to control. Hierarchy was necessary to ensure top-down direction of the authority and top-down direction was seen as a necessity of political control. Scale merely added necessity to necessity.

It could be argued however that the principle of hierarchy runs contrary to the traditions of professionalism. Professionalism rests upon the collective authority of the profession and, within that, stresses the autonomy of the individual professional. To an extent, therefore, professionalism sits uneasily with bureaucratic hierarchies. Managerial relationships must take account of the fact that they are between professionals. That can create tensions where the managerial actions of the superior are seen as interventions in the professionalism of the subordinate. Professionalism can lessen the detail of control within the hierarchy, but the hierarchy can also lessen the autonomy of the professional.

These three principles of functionalism, uniformity and hierarchy interact in support of each other. Hierarchy within clear functional divisions requires the adoption of uniform policies throughout the area of the authority. The principles are so much engrained in the working of the local authority that, for many within local government, they define the very nature and meaning of good administration. The principles give an appearance of stability and certainty to the working of the authority. They appear to ensure service delivery in accordance with rules laid down and good professional practice. They also appear to solve the problem of control in what is often a large-scale organization.

The Administrative Efficiency Movement

Even before 1935 there were many people concerned to improve the efficiency of local government. In the early 1930s Gibbon was arguing, in terms that are not unfamiliar in the 1980s, for efficiency audit, for greater use of unit costs and for measures of performance, which he called units of efficiency (Gibbon, 1931). There were also advocates of central purchasing and examples of its development (Steadman, 1932). Indeed throughout the period there have been many advocates of such particular measures to promote efficiency.

There have, however, also been those who were concerned, not so much with particular measures, but the need to review the whole organization in order to secure greater administrative efficiency. This movement for organizational review based on administrative efficiency drew much of its approach from advocates of classical theory such as Fayol and Urwick. Haynes in his *Organisation Theory and Local Government* described Fayol's concern for 'precise definition and rational division of functions in orderly hierarchical structures where authority is concentrated and clearly identified and where the emphasis is on order, clarity and simplicity' (Haynes, 1980, p. 5) and has argued that the classical theorists

> served to reinforce and lend theoretical legitimacy to principles of organisation which had already evolved and been applied as the logical solution to the problems of local government organisation. The legacy of classical theory was essentially a blueprint for the attainment of internal efficiency of operations, a means of imposing order and logic upon potential chaos. (Haynes, 1980, p. 15)

Urwick himself can be found writing about span of control and the functional principle in *Public Administration* in the 1930s (Urwick, 1935).

The administrative efficiency movement gained strength after successive decades despite Simon's criticism of many of its recommendations as contradictory (Simon, 1957). As Haynes implies, the administrative efficiency movement did not challenge the main organizing principles of local government. Functionalism, uniformity and hierarchy were accepted as the basis of administration. What the movement sought was the clear application of those principles. Thus the advocates of administrative efficiency argued for such changes as the merging of departments, the streamlining of committees, an emphasis on delegation and a stress on the administrative, rather than the legal, role of the Clerk. Organizational review was guided by the simple criteria of administrative efficiency rather than by the consideration of the role of the local authority as a unit of government. Very little changed in these prescriptions from the start of our period (and even before) until the Maud Report (Maud, 1967).

In 1929, for example, the Onslow Commission was receiving evidence that local authorities appoint 'an unnecessary number of committees'. Indeed the Association of Municipal Corporations were also of this opinion. Such a multiplicity of executive committees, it suggested, 'did not make for good and economic government and the Association therefore recommended that local authorities should scrutinize very carefully their practice and procedure in this matter with a view to a possible reduction in the number of executive committees and their separate staff, and the merger of all functions and duties which had a common relationship under one committee' (Onslow, 1929, pp. 114–15). Such a reorganization of committee structure was carried out in Edinburgh in the 1930s:

> In dealing with the question of the constitution of the new committees the Corporation took as their guiding principle the combining or grouping together of cognate administrative functions wherever that was possible, with the object of facilitating the discharge of the duties of members of the Council with due regard to economy of time and money. (Sir Andrew Grierson, quoted in G. M. Harris, 1939, p. 63)

Another theme of the movement was that the Clerk should have a special interest in administrative efficiency. This was taken up by the Hadow Committee on Qualification for Recruitment Training and Promotion of Local Government Officers which suggested that the Clerk was

> the chief administrative officer of the Council. The Council will look to him for advice on all major questions. He is the channel of their official correspondence, and responsible for the conduct of important negotiations on their behalf. The Clerk should co-ordinate the work of several departments, should keep in touch with the decisions of each of the committees, and should exercise a general supervision over all the work without interfering with heads of departments in strictly technical questions. (Hadow, 1934)

This emphasis on the Clerk as chief administrative officer gave expression to the classical theorists' emphasis on unity of direction. It was implicitly directed against the autonomy given by professionalism to the functional divisions of local government. Professionalism meant that there were limits to the extent to which the administrative efficiency movement could advocate the authority of the Clerk. The result was that, while the Conditions of Service for Town Clerks and District Council Clerks recommended in 1949 that the Clerk 'shall be the chief executive and administrative officer of the council' and 'shall be responsible for co-ordinating the whole work of the council', the recommended terms and conditions of chief officers did not define their relationship to the Clerk, but merely made them responsible to the council through the appropriate committees (Maud, 1967, p. 106).

This emphasis on the role of the Town Clerk as chief administrative officer was taken further in a Treasury Organization and

Methods Report on Coventry in 1953. This report, which received widespread attention in local government, suggested that the duties of this post should include:

(1) To take a continuing interest in the effectiveness and economy of all administrative arrangements throughout the Corporation.
(2) To ensure that administrative activities with which two or more departments are concerned are effectively co-ordinated.
(3) To act as Establishment Officer and to arrange for the provision of common office services where it is more economical to provide them on a centralised basis.
(4) To furnish an O and M service for all Departments.
(5) To maintain a broad view of the balance and effectiveness of arrangements made to carry out the policy laid down by the Council. (Greenwood and Stewart, 1974).

These proposals provided an organizational base to the concern for administrative efficiency; primary responsibility rested with the Clerk supported by Organization and Methods (O and M) units (later to be renamed management service units). Following this report there was a growth in the use of O and M in local government. By 1965, twenty-seven counties and twenty-five county boroughs employed O and M officers and a number of the others had employed O and M consultants (Harrison and Norton, 1967). The O and M officer of that day typified the movement to administrative efficiency in the application to organizational issues of the prescription of classical management. During the 1960s there were also a number of major organizational reviews carried out by local authorities. The appointment of Frank Harris, an executive with the Ford Motor Company, as Principal City Officer at Newcastle was followed by his report identifying thirty-four autonomous officers reporting to the council directly and the need to merge departments and clarify delegation (Elliott, 1971).

The most important illustrations of the administrative efficiency movement, however, can be found in the Report of the Maud Committee which was set up in 1964 to 'consider in the light of modern conditions how local government might best continue to attract and retain people (both elected representatives and principal officers) of the calibre necessary to ensure its maximum effectiveness' (Maud, 1967). The committee ignored its terms of reference, developing instead a critique of management practice in local government, along with a set of proposals for change. It argued, in effect, that the best way to attract and retain people of the calibre necessary was to change management practice. This provided the formal justification for a report whose focus on organizational structure represented the obvious interest and concern of the members of the committee.

The critique was clear:

> To allow any but the most trivial discretion to an officer is thought to be undemocratic. Hence the cumbrous machinery, the multiplication of committees, the clogged agendas, the long meetings, the neglect of matters of importance and the failure to make full use of the available controls over the administration and to develop others. (Maud, 1967, para. 72)

The committee referred to the Newcastle upon Tyne Review and argued that there 'are parallels in other major authorities. There is a danger that the number of departments and hence the number of professions will increase still further. The proliferation of departments and of principal officers make co-ordination of the authority's plans and their execution more difficult' (para. 104).

The analysis of the Maud Committee was straightforward. Large numbers of departments and committees were dangerous, as were long agendas. The driving force to administrative efficiency called for streamlining both the officer and the committee structure. From this analysis many of the recommendations clearly followed:

> We recommend that the number of committees of a local authority should be drastically reduced and that similar or related services should be grouped and allocated to one committee. (para. 169)

> We make no recommendations on the way in which departments should be grouped, although, in our view, their number can be reduced to half a dozen or so. (para. 227)

> While it is clear that the overall development and control of services should be the responsibility of members, in our view the day-to-day administration of services, the decisions in case work, the routine process of inspection and control should normally be

> the functions of the paid officers and not of the members. (para. 146)

The report also considered the position of the Clerk and recommended that the Clerk be recognized 'as head of the authority's paid service, and have authority over the other principal officers so far as this is necessary for the efficient management and execution of the authority's function' (para. 179).

Thus far the Maud Report can be seen as belonging clearly to the administrative efficiency movement. Because it was grounded in a movement already strong in local government and its recommendations fitted that movement, most of the recommendations described so far gained ready acceptance. Committees were reduced in numbers and departments were reorganized. There was a build-up in the authority of the Clerk who was renamed as Chief Executive Officer in certain instances.

There was, however, one further recommendation which was not so readily accepted. Thc Maud Committee recommended the setting up of a Management Board which would in effect take over the executive responsibility of committees. The Management Board's functions would be:

(a) To formulate the principal objectives of the authority and to present them together with plans to attain them to the council for consideration and decision.
(b) To review progress and assess results on behalf of the council.
(c) To maintain, on behalf of the council, an overall supervision of the organisation of the authority of its co-ordination and integration.
(d) To take decisions on behalf of the council which exceed the authority of the principal officers, and to recommend decisions to the council where authority has not been delegated to the management board.
(e) To be responsible for the presentation of business to the council subject always to the rights of members under standing orders. (Maud, 1967, para. 162)

Whereas some of the responsibilities of the Management Board belong to the principle of unity of direction and hierarchy accepted by the administrative efficiency movement, the recommendations about objectives and reviewing progress were an indication of emerging ideas and are related to other recommendations of the Maud Committee calling for 'a systematic approach to the process of management' (para. 187) involving the setting of objectives and the review of progress. These references were vitually ignored in the reports prepared by local authorities following the publication of the report (Greenwood, Norton and Stewart, 1969). They were unfamiliar ideas then, although they were soon to become much more familiar. In any event the recommendations of the Maud Report on the Management Board received the unique distinction of being rejected by every local authority in the country (Greenwood, Norton and Stewart, 1969). Its concentration of power in the hands of one committee and the apparent elimination of the executive role of other committees in favour of 'a representative and deliberative role' (Maud, 1967, para. 166) was just too much to accept.

Nevertheless, although the Management Board recommendation was rejected, part of the analysis which led to that recommendation was accepted. The committee had identified 'a lack of unity' in the working of local authorities. Its recommendations of a Management Board were directed towards achieving unity of direction. The analysis was widely accepted. The remedy was not. Instead, a large number of authorities set up policy committees with responsibilities for overseeing policies (Greenwood, Norton and Stewart, 1969). However, it is relatively easy to alter structures. It is far more difficult to establish the processes that give the structure meaning. Many more policy committees were set up to review policies than which actually managed to do so. Yet the fact that policy committees were established shows that the Maud Committee was not merely an expression of the administrative efficiency movement but also foreshadowed the corporate approach.

The Corporate Approach

From the Maud Report onwards, ideas which came to be known as the corporate approach came into prominence. The corporate approach differed from the administrative efficiency movement in that it did not take as its starting-point the existing activities of local authorities, but took instead as its starting-point the local authority as a unit of government. The corporate approach challenges the conception of local authorities, as a collection of bodies drawn together more by historical accident than for anything they have in common. In the words of the Bains Report it was argued:

> Local government is not, in our view, limited to the narrow provision of a series of services to the local community, though we do not intend in any way to suggest that these services are not important. It has within its purview the overall economic, cultural and physical well-being of that community, and for this reason its decisions impinge with increasing frequency upon the individual lives of its citizens.
>
> Because of this overall responsibility and because of the interrelationship of problems in the environment within which it is set, the traditional departmental attitude within much of local government must give way to a wider-ranging corporate outlook. (Bains, 1972, paras 2.10, 2.11)

This statement was an assertion that local authorities had a governmental role as well as a role in providing services of a predetermined pattern.

The corporate approach involved both a critique of existing practice and proposals for change. The critique had two main elements. As I have written elsewhere:

> The corporate approach was developed as a critique of the existing practices of management and policy-making in local government. That critique centred on the dominance of the department at the expense of the authority and on the emphasis on the administration of existing activities, rather than on the role of the local authority in the government of the locality.
>
> It was argued that, while any large organisation had to differentiate its activities, in local authorities differentiation was too strongly emphasised. It was not merely that local authorities were divided into departments but that the departments were matched by committees and their boundaries marked by a dominant profession. While the critique did not deny the need for departmentalism, it argued that differentiation must be matched by mechanisms capable of integrating the work of the authority and that mechanisms were inadequate in local government because they went little beyond administrative co-ordination.

> It was also argued that local authority management practices were geared to the maintenance of existing activities and that there was a tendency to treat those activities as ends in themselves rather than means to be reviewed in the light of the changing needs of local communities and of changing problems in the local environment. While recognising that the local authority must devote much of its energy to its established activities, this emphasis should not be carried to the point where review of activities was neglected. Local authorities needed to improve their capacity for review both of the needs and problems faced and of their activities. (J. D. Stewart, 1983, p. 166)

The challenge of the corporate approach went much deeper than the administrative efficiency movement. The organizing principles of functionalism were challenged directly by the corporate approach. There was a challenge to professionalism in that activities could not be justified merely by reference to professional prescriptions, but had to be reviewed in relation to needs and problems faced.

A distinction can be drawn between two aspects of the corporate approach. Corporate management deals with corporate working in the continuing functioning of the authorities while corporate planning focuses on the review of the activities of the authority and on the need to plan those on a corporate basis. The challenge of corporate planning, therefore, is deeper than that of corporate management.

The corporate approach came into prominence in the 1970s, but related ideas were already to be found in the Maud Report, as we have already seen. Much of the critique and the proposals that characterize the corporate approach have a longer history. Page, writing in 1936, made much the same criticism of policy-making in local government that was to be repeated nearly forty years later.

> Councils as such are unable to give proper regard to matters of general policy which is their chief function once the management of individual functions is left in the main to the committees . . . and the modern council tends to concern itself more with immediate detail than with general projected policy and the co-ordinating of the policies of which this general projected policy and the co-ordinating of the policies of which this general policy is made up. (Page, 1936, pp. 26–7)

Indeed the criticism goes even further back. Page quotes criticisms of Manchester City Council by the Webbs in 1899, who stressed that 'there was no central body to do that work of which the council was incapable – of deciding on the general larger problems of the city and of directing the work so that these problems were overcome by the concerted action of the Committee' (ibid., p. 27).

Page proposed that 'a small Co-ordination and Planning Committee should exist for the co-ordination of the general policies of all Committees. This infers [*sic*] the inability of the Council to perform this function. It also infers [*sic*] an important and moderately new departure, the preparation of long term policies and plans, taking into particular account the financial aspects of these plans and the development they involve'. He also advocated 'the more comprehensive grouping of Committee and departments' and the appointment of a City Administrator 'to co-ordinate the work of the departmental heads and to act as the Adviser of the Co-ordination and Planning Committee'. He stressed, in words anticipating the corporate approach that the 'whole of the system of co-ordination suggested is based on a conception of the local authority as a single entity to which the administration of various services is incidental' (ibid., pp. 407–8).

The ideas which Page represented were probably channelled after the war into the task of physical planning and redevelopment which dominated so many urban areas. The interest in the corporate approach and the forms of policy planning expressing that approach therefore tended to reflect an awareness of the limitation of physical planning in isolation as well as a growing concern for the problems created by departmental isolation in local government.

In the early 1970s a number of authorities took steps to introduce new systems of policy planning often based originally on planning, programming and budgeting systems, which attempted to relate expenditure in all departments to the objectives of the authority, (Greenwood and Stewart, 1974). Coventry introduced a policy planning process designed to produce a local policy plan which would 'define the Council's policies for the whole of the City, state in broad terms how the Council seeks to achieve their objectives, and its priorities for doing so, and it must have regard not simply for what the Council wishes to do, but also for its ability, as restricted by the resources available to it to carry out its plans' (ibid., pp. 239–40). With this innovation they were clearly moving beyond the perspective of the administrative efficiency movement, in which they had earlier played an important part in advancing.

It was however the Bains Report on *The New Local Authorities: Management and Structures* which marked the widespread formal acceptance of the corporate approach in local authorities. The emphasis of the Bains Report was more on management structure than on management process. That was the task it had been set up to carry out, and that was where the new local authorities wanted immediate help. The main recommendations of the Bains Committee, however, were designed to create structures to support the corporate approach within the local authority:

> We believe that the need for a corporate approach is beyond dispute if local government is to be efficient and effective. We recognise that there are widely differing views on how it can be achieved, but it will not be sufficient merely for this principle to be recognised. A framework must be built into the organisation of the various public services within which the idea can take root and develop. (Bains, 1972, para. 2.12)

The proposed structure was clearly based on a corporate approach and on a corporate approach that went beyond corporate management and towards corporate planning. The Bains Report proposed a Policy and Resources Committee, which would 'aid the Council in setting its objectives and priorities' (para. 4.14). At officer level the Bains Committee laid a parallel stress on a small management team of chief officers. The leader of the team would be a Chief Executive, who did not need a department of his own because he would work through and with the other departments. The Chief Executive would 'ensure that the resources and organisation of the authority are utilised effectively towards the attainment of the authority's objectives' and be 'the council's principal adviser on matters of general policy' (Bains, 1972, para. 5.13).

These recommendations were widely adopted and, despite changes in a number of authorities, the resulting structure of policy and resources committee, chief executives and management teams have become part of the accepted fabric of local government. There have been modifications. The Chief Executive without a department has become less common. The management team has in some authorities become less important than was originally expected. In some authorities the Chief Executive's role has developed as the main link between the political leadership and the officer structure.

The new organizational structures survived in part because they met needs created by the growing politicization described by John Gyford (Chapter 4). That politicization gave importance to the political leader, whose perspective can be regarded as corporate, as against the functional perspective that marks the role of the committee chairman. The policy and resources committee provided a clear role as chairman for the political leader. The Chief Executive position met political requirements for support for the leadership within the officer structure.

Whereas in the Bains Report the main emphasis was placed upon the Chief Executive working with the management team, in practice the management team has become less important to the Chief Executive. Rather than a key part of the corporate structure, the management team has tended to become a meeting of heads of the functional departments with the Chief Executive who is himself the main contact with the political leadership. As the management team has become less important as an expression of corporate management, the Chief Executive has tended to build up his own departmental base.

Furthermore, the era of financial constraint and cut-back has placed budgetary problems at the centre for both political and management concern. The corporate approach, far from being the planning of growth, has become the implementation of cut-back. The relationship between Chief Executive and Treasurer is therefore of critical importance. In some authorities Chief Executives have sought to redefine their role to emphasize their part in the budgetary process. In others there has been examination of the organization of the central services of finance, administration and personnel.

Broadly it can be said that following the Bains Report there has been some challenge to the domination of the functional principle, although it is still powerful. The role of Chief Executive has become established and has gained some support from growing politicization. The corporate approach has become dominated by the necessities of constraint. In so far as the functional principle has been challenged it may be more by the short-term pressures of cut-backs than by the challenges of longer-term policy planning.

The Bains Report itself gave little consideration to the processes of corporate planning. Nevertheless a significant number of local authorities after reorganization sought to develop corporate planning processes building on the early experiences of Coventry and elsewhere. These took different forms in different authorities. Many of the early attempts to develop corporate plans came to grief because those plans were too detailed, specifying not so much key policies, but the year by year detail of each activity. Those plans, based as they were on assumptions of growth, were challenged by financial constraint and cut-back. Corporate planning processes in so far as they have survived have tended to centre on immediate resources issues (Skelcher, 1982). Added to financial constraint, the introduction of block grant introduced new uncertainties and led to a shortening of the time horizon of decision-making. In situations of high uncertainty longer time perspectives are rejected (M. Davies *et al.*, 1983).

More recently, however, there has been a renewed interest in forms of forward planning. In part this reflects the emphasis given by the newly created Audit Commission to improved management systems. In addition, the cumulative effects of continued constraint have exposed problems requiring a longer-term perspective and a readiness to query, against that perspective, many existing activities. Many local authorities have responded to financial pressure with short-term solutions which, continued over a number of years, can create long-term problems. Thus in the short term, cuts in maintenance are often readily made, but in the long term the effects on property and roads can be serious. A decision made to undertake a capital project can pre-empt revenue resources, if those long-term revenue needs are not considered alongside capital projects.

Emerging forms of corporate planning can therefore be distinguished from the early development: they are more likely to be resource-centred; they will be more clearly related to political processes; and there may be an emphasis on priority setting. Selectivity of key issues and policies as areas of priority have thus proved more robust in uncertainty than comprehensive corporate plans (Skelcher, 1982; J. D. Stewart, 1983). In general it can be said that neither the aspirations of Harry Page in 1936 nor those of the advocates of the corporate approach in the 1970s have been realized. There have, however, been changes both in structure and processes. It is easy both to exaggerate and to underestimate their importance. Generally it can be said that there has been a growth in the corporate as opposed to the functional interest in local authorities, although the situation varies greatly between authorities. The emphasis has however been on a corporate concern for resources, more than for the needs and problems faced. The corporate approach as an idea nevertheless continues to challenge the functional. While a corporate perspective has been introduced, it is not sufficiently strong to replace the dominance of the functional principle. The corporate approach has also challenged the rigidity of managerial professionalism. The challenge was implicit rather than explicit, and was not likely to be effective. The challenge of the new politics has been greater.

The New Politics

The new challenge is a political challenge. Since 1974 the politics of local authorities have become more assertive. The change and the factors behind that change are described by John Gyford in Chapter 4. Councillors of both the left and the right (often, in the more volatile politics of recent years, coming into power more quickly, and concerned to bring about political change) have challenged not merely existing policies, but the very way of managing the authorities. The services provided are seen to be a product not merely of council decisions, but also of the organization of the local authority and in particular of managerial professionalism.

The new politics has brought an increase in full-time councillors, who may seek contact with staff at all levels of the organizations. Panels, working parties and review groups have been added on to the formal committee system as providing new settings. The political group has been recognized in an increasing number of authorities as part of the decision-making processes. The new politics has added new settings and new procedures which, while retaining the committee structure, have reduced it to a yet more formal role. The reality of the new politics can only with difficulty be encompassed in the old forms. While the nature of the changes varies with the politics of the authority, change is widespread.

Though the new politics has already challenged the forms of councillor–officer relations, there is a wider challenge to managerial professionalism and to the dominant organizing principles of functionalism, uniformity and hierarchy. That challenge involves both a critique and proposals for change.

The critique focuses on the processes by which managerial professionalism defines the needs of client and customer, fitting them to the requirements of the organization. Professional and bureaucratic authority reinforce organizational dominance in the relationship between authority and client or customer. The principle of functionalism and the professionalism that supports it are regarded as categorizing problems in accordance with organizational divisions rather than people's own perceived needs. The principle of uniformity prevents responsiveness to individual needs or local problems. Hierarchy means that the fieldworker is not accountable to client or customer but controlled by those above him, who are both organizationally and geographically remote from the field.

Donnison has charted the change to which the new politics is now responding.

> Since the beginning of the twentieth century, social reformers in many fields had sought to replace local charities, the local poor law and local friendly societies with nationwide services. These services were then pressed to adopt increasingly similar practices throughout the country. Official guidance about pupil–teacher ratios, about prescriptions per doctor and per patient, about points systems for the allocation of council housing and about codes for the distribution of social security benefits were all designed to produce greater uniformity. Radicals helped to standardise things. Variety, they argued, created 'anomalies', and anomalies were 'indefensible'. Always alert to seize upon these anomalies as precedents for reform, they compelled government to adopt an increasingly uniform mould in self-defence.
>
> The achievements of this heroic age could perhaps only have been attained by a set of fairly single-minded, centralised bureaucracies which focused and confined the energies of politicians and officials within the frontiers of clearly distinguished departments of central and local government. A functional, service-oriented system was created to win the 'numbers games' of these years: to build the houses, to clear the slums, to get the new pensions paid, to ensure there would be a place in school for every five-year-old, to deploy hospital consultants in every region, to build more miles of motorway and to bring down the numbers on the waiting lists. The heroes of these games – the chairmen of housing and education committees, their chief supporters and senior officials – too

> often became detached from the people on whose behalf they laboured. They thought of these people as tenants, pupils, patients, social work clients, losing sight of the more complex human reality behind the flood of committee papers and the smoke of bureaucratic battles. (Donnison, 1983, p. 2)

The critique of managerial professionalism is in part shared by radical left and radical right. Both challenge the assumptive world of local government in which client need is defined organizationally according to established professional prescriptions. But while the critique is to a degree shared, the response differs. To the radical right the response should be sovereignty of the consumer in the market-place (Adam Smith Institute, 1983). To the radical left the response must be based on responsibility to local communities (Blunkett and Green, 1983). The proposals associated with both responses are wide ranging. From the radical right one finds proposals for privatization and the withdrawal of government, use of fees and charges, use of volunteers, voucher schemes and private contracting. From the radical left one finds proposals for neighbourhood offices, decentralized political structures and use of community groups.

All these proposals challenge the standard modes of service delivery. In the past when local authorities have been given a function, it has been assumed without question that the service would be directly provided by the local authority through a hierarchy responsible to committee and council. Managerial professionalism was unchallenged. Now that challenge is there and while some of the proposals have a longer history, this is a development of recent years.

It is so recent that although a few classic cases are well known – for example the privatization of refuse collection in Southend – actual changes are limited. Limited experiments in decentralization have taken place, and again classic cases such as the Walsall neighbourhood housing offices are much commented on.

The new politics challenges the dominant organizing principles of managerial professionalism. Those principles, however, are not easily challenged. They are part of the very working of local government. There is a compulsion in the organizing principles. The large-scale delivery of public services may appear to require functionalism, hierarchy and uniformity. The criticisms of managerial professionalism are also its virtue. The apparent rigidity of uniformity may be the guarantee of fair and equal service. The unresponsiveness of hierarchy may enforce impartiality. The apparent fragmentation of functionalism may avoid the irrelevance of the overgeneralized approach.

Consequently, the challenge of the new politics may neither destroy managerial professionalism nor overcome the organizing principles. Nevertheless, it may create countervailing forces in the organization which limit the dominance of those principles. These developments may, however, belong more to the future than to the past.

Conclusions

As local authorities have experienced a considerable growth in scale, over the last fifty years, it was probably inevitable that the organizational problems involved would be dealt with in accordance with well-established organizing principles. The building blocks of the functional divisions have grown in scale, the organizational hierarchies have lengthened, the uniformities of policy have been extended over wider and often disparate geographical areas. As local authorities have grown in scale, the principles of managerial professionalism seem more necessary, but the problems associated with them become more obvious. It is not surprising that the challenges become greater.

The strength of managerial professionalism is clear. Professionalism brings specialist skills and established standards. The dominant organizing principles secure regularity in the working of local authorities. Larger-scale delivery of services is secured. The organizing principles ensure stability as they ensure uniformity. They provide a mechanism for formal political control through the hierarchies. Managerial professionalism supports the local authority's role in service delivery. Its emphasis is upon the maintenance of those services on well-established patterns. It sustains the view of the local authority as the provider of a set of predetermined activities.

The local authority is both a provider of services and a unit of government. As a provider of services, the emphasis is upon maintenance. As a unit of government the emphasis is upon response to a changing environment and upon change directed by political decision. The organizational dilemma for local government is how to combine in its structure and process the service requirements of maintenance, and the governmental requirements of change.

The organizing principles that support maintenance can limit change. Functionalism, supported as it is by the assumptive world of professions focused on particular functions, provides a clear focus on the established tasks that have to be performed, but can reduce the capacity of the authority to deal with new and emerging issues in its environment. Uniformity ensures the delivery of services, but limits both responsiveness to a diversity of needs and demands and can limit the learning that comes more easily from difference than from sameness. The hierarchies of managerial professionalism have the rigidity of established tasks, but lack the flexibility for learning and for the adaptability that goes beyond those tasks.

The principles of managerial professionalism have been challenged, not by the administrative efficiency movement, which reinforced its principles, but first by the corporate approach and now by the new politics. The corporate approach draws its force from the conception of the local authority as a unit of government and challenged the principle of functionalism. The challenge of the new politics is directed at all three organizing principles.

The organizing principles are not easily overthrown because they remain deep in the very working of local government. Functionalism, uniformity and hierarchy define the accepted concept of good administration. And the local authority remains a provider of services as well as a unit of government. It has to encompass maintenance and also encompass change both in the services and their governing. That is the dilemma.

The organizing principles are likely to remain important in the working of the local authority for so long as it plays a role in the large-scale delivery of services. The response to challenge may be that the local authority can also develop a response based on other principles, just as the corporate approach has built a corporate perspective to complement and to challenge, but not to replace, a functional perspective. Both hierarchy and uniformity may remain organizational necessities for certain roles, but not for all. The local authorities may encompass organizational difference.

The response remains uncertain – above all to the emerging

new politics. In the meantime it is probably impossible to say whether there has been progress or regress in the functioning and management of local government, because there have been both. The strengths of the organizing principles are also their weakness.

6

Administrative Law, Local Government and the Courts

MARTIN LOUGHLIN

> The purpose of judicial control is not ... to prevent authorities from acting, but to induce them to act wisely and well.... The general conclusions to be drawn from the past century are by no means favourable to judicial control. To some extent the defects are inherent in judicial control.... However, when we consider the actual practice of judicial control, apart from its inherent defects, we cannot feel satisfied. (pp. 427–8)
>
> The result of all these methods of administrative control is to make Central Government an essential factor in what is still called 'local government'.... It is true that we have elected local authorities exercising a discretion according to the opinions which meet the approval of their own electorate. It is true also that they can do a great deal, within their powers, to improve the health and happiness of their constituents. But they are rigidly restricted to the powers conferred on them by Parliament; their organisation and their proceedings are determined by statutes ... and above all they are controlled more or less closely in all their activities by organs of the Central Government. (p. 450)
>
> 'Central control' must ... involve, not a meticulous investigation of the details of local administration, but a general control in the national interest of divergent or contradictory local policies. (p. 454)
>
> Ivor Jennings in *A Century of Municipal Progress*

The history of public law thought in this country is marked by a tension between the ideas of administration and law. For some a sharp distinction must be made between administration and law.

Administration is seen as a method of achieving policy objectives by the most effective means and is undertaken within the constraints imposed by law. Law is viewed as a body of norms which may be applied, in accordance with an authoritative technique, in order to determine disputes which arise in the process of administration. By way of contrast, however, others emphasize the similarities rather than the differences between administration and law. Indeed, by emphasizing the purposive nature of both activities and the open texture of law, it is even suggested that there is no fundamental distinction between the administrative process and the judicial process. Clearly the view which is taken of the relationship between administration and law colours the attitude which one takes to not only the relationship between the administration and the courts, but also the nature and significance of administrative law.

Nowhere has this tension between administration and law been more evident than in the area of local government. Local government functions are undertaken by elected local councils. In law, local councils are statutory corporations and, as local authorities, are given powers and duties to carry out various functions. However, these elected local councils also constitute an important tier of *government*: the functions vested in local authorities are not entirely random; in exercising these responsibilities they are given the capacity for 'local choice'; and the combination of structure and functions is such that collectively the institution is more than the mere aggregation of its individual statutory functions.

This is recognized in law by the breadth of discretion given to local authorities by statute. Nevertheless it has often been felt that the courts, in reviewing the decisions of local authorities, have failed implicitly to recognize their status as governmental institutions. In the infamous case of *Roberts* v. *Hopwood* in 1925, for example, Lord Sumner apparently denied that elected members of local authorities ought 'to be guided by their personal opinions on political, economic or social questions in administering the funds which they derive from levying rates'.

This tension between administration and law is therefore particularly evident in the sphere of local government. The editors of *Municipal Progress* themselves had a distinctive view on this issue which influenced the views expressed in that volume and their writing on public law in general. In order to evaluate the significance of legal developments affecting local government since 1935, therefore, we must first examine the source and nature of this tension.

Normativism and Functionalism in Public Law Thought

The work of Jennings, Laski and Robson should be seen essentially as a response to the influence of A. V. Dicey. Dicey's *Introduction to the Study of the Law of the Constitution* (1885) was 'the first to apply the juridical method to English public law' (Jennings, 1933, p. x) and has been very influential. Dicey was writing during a period of great political and economic change, in which the range and the functions of the state expanded enormously and legislation was used explicitly as a means of changing social conditions. This attempt to use law as a tool of social engineering resulted in the increased use of the regulatory form of law, which vested discretionary powers in administrative agencies, and led to the politicization of law. These developments therefore presented a challenge to the traditional authority of law. It is within this context that Dicey's work should be examined.

The traditional authority of law rested on the fact that the law consisted largely of customary norms which therefore provided the basis for a relatively impartial method of resolving disputes. Social legislation challenged this traditional impartiality of law, which itself underpinned the idea of the autonomy of law, since such legislation was manifestly political in origin. Social legislation also challenged the idea of the rationality of law because, far from establishing clear standards of behaviour, by cutting across traditional boundaries between public and private and by interfering with market expectations, it generated disputes on a larger scale without providing authoritative norms for resolving them.

Dicey responded to this challenge not by providing a detailed account of the law of the constitution as it existed in 1885 but by seeking to lay bare the fundamental legal principles which underpinned the British constitution. Two basic principles were discovered; the sovereignty of Parliament and the rule of law. It is with the latter that we shall be most concerned. For Dicey, the concept of the rule of law was founded on the principles of equality before the law and the supremacy of the ordinary law of the land. From these principles he derived a third; that no special legal status was accorded to the state, which was subject to the ordinary law as applied in the ordinary courts. That is, Britain had no special system of administrative law.

Behind every concept in public law, however, lies a theory of power and the terms on which it may be exercised. Dicey's theory may be termed a *normativist* theory of public law. He sought to protect the authority of law from the challenge posed by the growth of social legislation by undermining the authority of such legislation. The autonomy and rationality of law was thus defended by incorporating within his concept of the rule of law a particular concept of law. Law is viewed as a body of specific norms in accordance with which disputes can be impartially resolved. Hence, the rule of law, in Dicey's view, 'is contrasted with every system of government based on the exercise by persons in authority of wide, arbitrary, or discretionary powers' (Dicey, 1885, p. 184). Since, as we have seen, social legislation tended to generate disputes without providing authoritative norms for resolving them such legislation was viewed with scepticism and ultimately considered inimical to the concept of the rule of law. Dicey's normativist theory of public law consequently incorporates a sharp distinction between administration and law.

Dicey's views dominated public law thought during the half-century to 1935 as Jennings himself acknowledged: '[Dicey's] influence upon the development of public law in England has been immense. His authority to-day is greater than that of any other public lawyer' (Jennings, 1935b, p. 133). After 1925, however, Dicey's views were subjected to rigorous examination and criticism by a group of young public lawyers and political scientists led by Jennings, Laski and Robson. During this period they set about the task of demolishing the normativist theory of public law. In 1928 Robson published *Justice and Administrative Law* in an attempt 'to dispel the illusion held by all the leading lawyers, politicians, civil servants and academics who had been brought up on Dicey's *Law of the Constitution* that in Britain there was no administrative law' (Robson, 1979, p. 107). In 1933 Jennings published *The Law and the Constitution* which critically examined the constitutional principles articulated by Dicey, was especially critical of his concept of the rule of law, and sought to reorient the focus of public law away from Dicey's concern with individual rights towards an examination of the powers and functions of public authorities. And during this period the normativist view of adjudication was criticized by emphasizing the discretionary power of courts and the philosophical, cultural or political biases displayed by judges exercising that power (for example, Laski, 1926, 1932; Jennings, 1936).

Their work provides us with a systematic critique of the normativist theory and an alternative approach which is based on what may be termed a *functionalist* theory of public law. This functionalist theory attempts to break down the dichotomy erected by the normativists between administration/discretion/policy and law/rules/rights. The similarities between administration and law are explored, especially by emphasizing that creative judgement and a unity of purpose underlies both activities. It is this focus on *purpose* which is definitive of their approach. Accordingly, the primary function of public law is to devise institutions and procedures in order to establish a framework within which public services can be most efficiently and equitably provided.

The period 1925–35 was also a period of intense conflict between normativist and functionalist approaches to administrative law. In 1929 Lord Hewart, the Lord Chief Justice, wrote *The New Despotism*. Working within a Diceyan normativist framework, his thesis was that the recent growth in power of the administration had had the effect of placing it beyond the control of both Parliament and the courts. This led, more or less directly, to the establishment of the Committee on Ministers' Powers, of which Laski was a member. Its report (Donoughmore, 1932), which generated much discussion (for example, Jennings, 1932; Robson, 1932a), adopted neither an extreme normativist nor systematic functionalist (Robson, 1951, pp. 426–33) analysis and adopted the pragmatic middle way of incremental reform.

The debates of the previous decades undoubtedly influence the approach taken in *Municipal Progress*. Thus Jennings goes out of his way to defend the administrative powers of central departments and local authorities from Hewart's criticisms (Jennings, 1935a, p. 442) and criticizes the normativism of the judiciary who have tended 'to delight in interpreting statutes in an obstructive manner (ibid., p. 428). His functionalist approach also determines his conception of progress. Progress, it appears, stems from a shift in emphasis during the nineteenth century from judicial control to administrative control over local authorities, and through the emergence in the early twentieth century of a comprehensive system of administrative supervision by central government.

The key themes in this analysis, many of which are evident in the quotations of Jennings which open this chapter, are first, the restrictiveness of the *ultra vires* doctrine; second, the inadequacies or limitations of the courts as supervisory bodies; third, the potential superiority of administrative methods of review; fourth, the need to entrust local authorities with sufficient powers to enable them effectively to provide public services which meet the needs of their constituencies; and finally, the need for an efficient system of supervision by central government in order to resolve conflicts between divergent or contradictory local policies.

The history of legal developments affecting local government may thus be evaluated in terms of the competitive influences of the normativist and functionalist approaches to administrative law. The three issues which seem most significant are first, the relationship between local authorities and the courts; second, the need to establish 'internal' standards of good administration which structure and check the exercise of local decision-making; and third, the need to entrust local authorities with sufficient powers to enable them effectively to provide public services which meet the needs of their constituencies. These issues will be considered in turn.

Local Authorities and the Courts

In the year following the publication of *Municipal Progress* Jennings published an important article on the relationship between local authorities, central government and the courts in the implementation of housing legislation (Jennings, 1936). In this he argued that, especially between 1928 and 1935, the result of judicial intervention was to frustrate the effective implementation of legislation designed to improve housing conditions. This he suggested was because of a judicial bias against social legislation. In effect he argued that this arose from the adoption of a normativist conception of public law, which resulted in the failure of the courts to take a purposive approach towards interpreting the legislation; the imposition of a legalistic structure on administrative decision-making; and a general spirit of hostility towards the discretionary powers vested in the administration.

It is this experience which led Jennings (1935a, p. 429) and Robson (1928, ch. 6) to propose that a specialist court was needed to deal with administrative law. Jennings certainly saw the need for some form of review of administrative decision-making: 'Private interests must give way to the public good; but the public good is ultimately injured by acts which though apparently in its favour, lead to individual injustices' (Jennings, 1936, p. 454). But he felt that the normativist orientation of the courts, together with the fact that they 'usually have had no experience of the problems which local government have had to face' (Jennings, 1935a, p. 428), led them to favour private interests over the public good.

After 1935, however, this spirit of hostility to local authority action waned (see Robson, 1959). Indeed the period from 1935 to 1965 may be characterized as one of judicial acquiescence to administrative decision-making. Thus Professor Griffith states that it is

> really only in the course of the 1930s that the courts began to accept these powers with any grace. In the Second World War the acceptance began to border on enthusiasm. This can no doubt be explained by the particular national emergency of those years. But it was perhaps surprising that this attitude continued when the courts were faced with the legislation passed to deal with post-war problems. (Griffith, 1959, p. 116)

According to Griffith the judges seemed on occasion 'to be leaning over backwards almost to the point of falling off the Bench to avoid the appearance of hostility' (ibid., p. 120).

The spirit in which judicial review of local authority action was undertaken during this period is perhaps best captured by Lord Greene MR's judgement in *Associated Provincial Picture Houses Ltd* v. *Wednesbury Corporation* in 1948. In this the Master of the Rolls set out the principles of *ultra vires* as they applied to local authorities and emphasized in particular the fact that, since local authorities are entrusted by Parliament with a decision within their knowledge and experience, they can be best trusted to deal with it and, therefore, that the court is not concerned with the reasonableness *per se* of that decision, since that ultimately would involve them in substituting their decision for that of the authority.

How do we explain this period of judicial self-limitation? Given the fact of the welfare state, the courts were placed in some difficulty. The existence of administrative powers seemed to require that they develop a new mode of adjudication (Chayes, 1976) and a new jurisprudence. Yet this is precisely what the normativist theory could not countenance since the theory implies the existence of a single indivisible system of law which does not recognize the special position of the state or the administration. They had therefore either to impose an unrealistic legalistic framework on administrative decision-making (to protect a perceived challenge to the rule of law) or retreat from activist judicial review on the ground that the minister was responsible to Parliament for administrative decisions or, in the case of local authorities, that 'representatives may be trusted to understand their own requirements better than judges' (*Kruse* v. *Johnson*). Once the courts realized that these administrative powers were a permanent feature of the modern state it is hardly surprising that they adopted the latter approach.

The irony is that Dicey's criticism of *droit administratif* arose essentially from the belief that the separate administrative jurisdiction in France gave a special immunity for official acts. In the result, by the 1950s, even the judiciary were aware that 'so far from granting privileges and immunities to public authorities, the French administrative courts keep them in order and exert a supervision and control over them which is more complete than anything we have here' (Denning, 1949, p 80).

Although some judges felt that the courts no longer had 'the strength to provide any satisfactory solution to the problem of keeping the executive ... under proper control' (Devlin, 1956, p. 14), beginning in the mid-1960s there commenced a new wave of judicial activism. Thus, while in 1964 Lord Reid admitted that 'we do not have a developed system of administrative law' (*Ridge* v. *Baldwin*, 1964), by 1982 Lord Diplock felt able to state that 'the progress towards a comprehensive system of administrative law . . . I regard as having been the greatest achievement of the English courts in my judicial lifetime' (*R.* v. *Inland Revenue Commissioners, ex parte National Federation of Self-Employed*).

Local authority action has certainly not been left untouched by this new wave of judicial activism but the main result has been to apply the basic concepts and the old remedies with greater flexibility and the main innovations have been in judicial review of ministerial action and the decisions of non-departmental public and quasi-public bodies. Nevertheless the greatest failure remains the failure of the courts to develop a rigorous and realistic administrative law jurisprudence.

Again, nowhere is this more clearly evident than in relation to local government. The principles enunciated by Lord Greene in the *Wednesbury* case remain the most detailed elaboration of the principles of judicial review of local authority action. Nevertheless, these principles are used essentially as a mask to conceal the real basis for judicial decisions and it seems true to say that 'the courts have been able to use the *Wednesbury* formula as a convenient rationalisation for varying policy decisions' (Elliott, 1980, p. 583). Thus, if a court chooses to intervene it will tend to utilise the 'irrelevant considerations' aspect of the formula, whereas if it declines it will emphasize those parts of the judgment dealing with 'unreasonableness'. Recently this fact has been recognized by Ormrod LJ who suggested that Lord Greene's judgment is 'coming to be regarded as a proposition of law in itself' and is 'indiscriminately used to express different and contradictory ideas' (*Pickwell* v. *Camden LBC*, 1983).

Consequently Jennings's criticisms of the mid-1930s are of equal applicability today. The courts still fail to recognize local authorities as institutions of government: 'the subsidising of particular classes of society is . . . a matter for Parliament, and for Parliament alone' (Vaisey J in *Prescott* v. *Birmingham Corporation* in 1955). They still invoke anachronistic doctrines, such as that of a fiduciary duty owed by a local authority to their ratepayers (for example, *Bromley LBC* v. *GLC* in 1982) or 'the well known principle that a statute should not be held to take away private rights of property without compensation unless the intention to do so is expressed in clear and unambiguous words' (*Hall & Co. Ltd* v. *Shoreham-by-Sea UDC* in 1964). As a result, they have gone no further in developing, for example, an appropriate conception of private property rights under modern regulatory regimes (see Loughlin, 1985).

In 1931 Jennings (1931, p. 3) wrote that 'it is indeed a remarkable fact that so often a decision of a court acts as a spanner in the middle of delicate machinery'. What is even more remarkable is that the same is true today. This is clearly illustrated by the case of *Bromley LBC* v. *GLC*. Here, the House of Lords quashed a supplementary precept issued by the GLC to finance a fares reduction on London's public transport system on the ground that it was beyond the powers of the GLC laid down in the Transport (London) Act 1969. They did so basically by construing a technical accounting provision, section 7 (3) of the 1969 Act, as incorporating a limitation on the power of the GLC to make revenue grants, and in the process overturned more than a decade of conventional wisdom and practice in the area of public transport policy. It required a couple of lower court judgements (*R* v. *Merseyside CC, ex parte Great Universal*

Stores Ltd in 1982 and *R* v. *London Transport Executive, ex parte GLC* in 1983) and a new Act of Parliament (Transport Act 1983, section 2) to re-establish a coherent legal framework in this area.

Furthermore, in that case Lord Diplock, although recognizing the majority's errors of statutory construction, nevertheless concurred in the result on the ground that, since the impact of the GLC's decision resulted in loss of rate support grant, the rate burden of this decision was such that the GLC were in breach of their fiduciary duty to ratepayers. This judgement had even wider ramifications. It is clear that Lord Diplock did not understand the nature of the block grant system introduced by the Local Government, Planning and Land Act 1980 since the strict logic of his judgement resulted in absurdity (see Loughlin, 1983, pp. 45–7). Nevertheless, it led to attempted legal challenges to entire budgets of local authorities and generated a great deal of business at the local government bar where counsel were constantly asked to advise on the legality of a great number of local authority policy decisions, ranging from the level of council house rents to the price of school meals.

Thus, the courts often seem incapable even of articulating the issues properly, let alone adjudicating on them in an appropriate manner. However, these issues are not often discussed widely owing to the normativism which still pervades administrative law scholarship. Normativism is based on a simple model in relation to such issues: administrative discretion is suspect; the greater the incidence of judicial review the better; judicial decision-making is correct, or at least is not a subject-matter for criticism (for example, Wade, 1982).

Nevertheless, since the late 1970s there have been several developments which suggest that a functionalist approach is gaining in support in the courts. In grappling with the issue of the liability of local authorities for negligence in the exercise of statutory powers, for example, the House of Lords in *Anns* v. *Merton LBC* in 1977 recognized that the principles apply differently to public bodies discharging functions under statute since their powers are definable in terms of public rather than private law. Since then this distinction between public and private law has been developed and the seeds have been sown for the establishment of a special system of administrative liability.

Alongside this development has been the reform of administrative law remedies with the establishment in 1977 of the application for judicial review. The importance of this reform has been increased in two ways. First, because since *O'Reilly* v. *Mackman* (1983) it seems that the application for judicial review is becoming the exclusive remedy for challenging administrative action (cf. *Wandsworth LBC* v. *Winder*, 1984). Secondly, because, as a result of administrative changes, High Court judges with special expertise in administrative law have been assigned to hear all cases of judicial review (Blom-Cooper, 1982). Consequently, recent changes have led to the establishment *de facto* of a specialist administrative division of the High Court.

These developments have been criticized from a normativist perspective. It has been argued that these developments breach the Diceyan principle of equality before the law, that the procedural reforms will lead only to barren jurisdictional disputes and that the public/private distinction has resulted in the creation of special immunities for the administration and 'is nothing more than an attempt by the judiciary to conceal political issues behind a formalist facade and to shield from public criticism some highly executive-minded decisions' (Harlow, 1980). Jurisdictional disputes certainly have resulted (for example, Sunkin, 1983). But others have argued that the recent procedural changes could provide a framework 'in which a new jurisprudence in administrative law can flourish' (Blom-Cooper, 1982, p. 260) and that '"executive" thinking by the judiciary is unlikely to be affected by the formal abolition of the category of public law' and therefore, instead of pretending that there is no distinction between a person and the state, these developments should cause us 'to think more positively about the relationship between the individual and the state' (Samuel, 1983).

Nevertheless, these developments are rather different from the specialist and autonomous jurisdiction which was proposed by Robson (see 1951, pp. 426–33) in his submission to the Committee on Ministers' Powers. Interestingly, in his evidence before the Franks Committee in 1957 Robson was very critical of such piecemeal reform (Franks Report, Memo. of Evidence, p. 492):

> The Administrative Appeal Tribunal should be separate from the High Court of Justice. Otherwise it will not be sufficiently free from the tremendous weight of tradition and precedent which bears so heavily on the ordinary courts. Merely to establish an Administrative Division of the High Court would be a reform in name only rather than in substance. It would be dominated by the

> cumbersome procedure, the narrow methods of interpreting statutes, the common law rules of evidence, the reliance on existing remedies and forms of action, and exhorbitant costliness, which characterise the British legal system.

Robson at least would not have been much impressed by these reforms; and, if some of his criticisms are not as acute today as they may have been in 1957, other doubts about the reforms exist (see Craig, 1983, pp. 498–517). Consequently, in terms of the relationship between local authorities and the courts little has changed over the past fifty years and the period ends much as it opened – with the working through of various incremental reforms to the system of administrative law.

Administrative Controls over Local Authority Action

For many functionalists, however, the recent developments in judicial review will be only of marginal interest since they have all but abandoned courts as a subject of serious study. The reason for this is that, with the growth in government, courts have become peripheral to the challenge of ensuring that administrative powers are exercised efficiently and equitably. For the functionalists the real challenge is the establishment of administrative procedures which encapsulate the ideal of good administration.

Thus, although local authorities are given a great deal of discretion by Parliament, in exercising their powers they are obliged to comply with an array of administrative procedures and are subject to a range of checks. Before making many decisions, for example, local authorities are obliged to comply with statutory procedures which make provision for notice, consultation and the like. Furthermore central departments and their quasi-independent offshoots possess a host of powers of appeal, approval, inspection, appointment, audit and default which structure local authority decision-making. Many functionalists therefore view their role as advising on the design of appropriate decision-making processes. Robson, for example, appeared to do so, which may well explain his mid-career switch from law to public administration.

Over the last fifty years, as the structure, functions and role of local government has changed, so too has the nature of administrative controls over decision-making. Periodically central departments have shed themselves of detailed controls over local decision-making. In the field of planning alternatives to the public local inquiry, such as the examination-in-public in the context of structure plans and the written representation procedure for appeals against planning permission, have been fashioned. In 1972 the role of the direct auditor in relation to surcharge for unlawful expenditure was changed from that of an adjudicator to a quasi-prosecutor and in 1982 audit reforms were introduced with the aim of 'improving economy, efficiency and effectiveness in the provision of local authority services' (Local Government Finance Act 1982). As the scale of local bureaucracies increased, the ombudsman technique was adapted and a Commission for Local Administration was established in 1974 to investigate allegations of maladministration made against local councils.

Nevertheless, local authorities possess a great deal of discretion. With the growth of council housing, for example, they have become the major landlords in the country and have been given more or less unfettered discretion in the management of their housing stock. Certainly, the law did not intrude into the relationship between council tenant and local authority landlord. 'It is to my mind one of the important duties of management that the local body should be able to pick and choose their tenants at their will', said Lord Porter in *Shelley* v. *LCC* (1949), adding that it 'is true that an ordinary private landlord cannot do so, but local authorities who have wider duties laid on them may well be expected to exercise their powers with discretion'. That is, local authorities were given a free hand because they could be expected to act as model landlords. Unfortunately, this did not always prove to be the case (National Consumer Council, 1976; Housing Research Group, 1981).

Similarly with planning control:

> The history of town and country planning law is the history of the extension and increase of power over the urban environment and the decrease of meaningful control over that power. For every ill from which urban society suffers the same remedy is proposed; increase the power and discretion available to administrators and planners and then all will be well. (McAuslan, 1974, p. 138)

Studies demonstrated that the planners did not always know best (Dennis, 1970; J. G. Davies, 1972). Consequently the main opposition to the clearance schemes of the 1960s often came not from the slum landlords, as in the 1930s (Jennings, 1936), but from the residents themselves.

In short, by the 1970s there was widespread support for the view that local authorities 'have ceased to be a method, indeed they have become part of the problem, of finding democratic ways of living' (Hampton, 1971). Clearly, given the revelations of the Poulson Affair, it was not sufficient merely to have faith in the goodness of councillors (Gillard and Tomkinson, 1980). Given both the management practices of local authorities (see Stewart, Chapter 5) and the structural constraints under which they operated (CDP, 1977) many believed that local government was primarily a tool of social control (Cockburn, 1977). Furthermore, the functionalists, to the extent that they had too readily adopted a dirigistic approach in criticizing the orthodoxy of normativism, could not be entirely dissociated from such criticism.

The result was that once again the issue of administrative discretion emerged as a key political issue. The radical functionalist response was to challenge bureaucratic structures and priorities in local authorities to ensure that they served the needs of the putative beneficiaries of local services. The challenge was seen primarily in terms of finding effective methods of public participation in local government decision-making (for example, McAuslan, 1981). More recently, it has taken the form of reorganizing the method of service delivery through various decentralization schemes (Fudge, 1984).

However, beginning in the late 1970s a normativist counter-challenge emerged, which aimed to check administrative discretion by providing particular beneficiaries of local services with certain individual rights. This initiative is seen most clearly in the field of housing where in the Housing Act 1980 council tenants were given security of tenure similar to private sector tenants and also a right to purchase the freehold of their dwelling. More recently, in the Housing and Building Control Act 1984, tenants have been given a 'self-help' right to repair their dwellings. More generally, individual means-tested housing benefit is replacing collective rent-pooling arrangements as the primary means of subsidizing the rent of council tenants. Similar developments may also be seen in other areas such as education, where by virtue of the Education Act 1980, parents were given rights to challenge the allocation of their children to a particular school.

In this form the counter-challenge may be viewed as a normativist attempt to establish standards of good administration. However, this development may also be seen as one element within a broader challenge which moves beyond the issue of how local services may best be provided to embrace the question of whether the local authority should be providing the services at all. It is therefore to this issue that we now turn.

The Legal Foundations of Local Authority Action

Jennings assumes that a primary function of administrative law is to ensure that the legal framework through which local authorities operate enables them to provide public services effectively. The key factors are the breadth of statutory powers and the extent to which central controls restrict local authorities. These issues are not dealt with in detail in *Municipal Progress* but Jennings (1931, pp. 22–3) had earlier expressed his views:

> The fact that local authorities are strictly limited by the rules of law and that the central departments are generally able to prevent the acquisition of new powers with which they disagree, the close financial control which is exercised by means of control over loans, control through the audit, and control through the new grant system, all of these have given the Ministry of Health in particular a predominant power in local government.
>
> In many respects this close control has been to the great advantage of the public. On the one hand it has stimulated, nay, compelled sluggish authorities to provide the amenities which local opinion has been too weak to demand. On the other hand it has curbed the extravagance of more impetuous bodies.
>
> Has it brought no disadvantage? I think it has. It has, in particular, stifled local initiative. Its natural tendency has been to create the belief that stimulus must come from the centre. Local authorities have only those powers which Parliament has given them. Parliament today means the central departments. . . . It follows that the main developments of local administration are the consequence of the initiative of the central departments.

Jennings felt that this situation was unsatisfactory and, believing that the problems of local government were inextricably linked to problems at the centre, proposed that local government should be given greater freedom through a more general scheme for the decentralization of government (ibid., pp. 25–8). This is a sophisticated analysis. Unlike Robson (1928, pp. 195–9), Jennings did not put the blame for this state of affairs entirely on the doctrine of *ultra vires*. Nevertheless, during the last fifty years that doctrine has often been blamed for many of the problems and failures of local government, as the views of the Maud Committee (1967, para. 283) demonstrate:

> Our view is that ultra vires as it operates at present has a deleterious effect on local government because of the narrowness of the legislation governing local authorities' activities. The specific nature of legislation discourages enterprise, handicaps development, robs the community of services which the local authority might render, and encourages too rigorous oversight by central government. It contributes to the excessive concern over legalities and fosters the idea that the Clerk should be a lawyer.

Undoubtedly, as Laski himself would have agreed (see Laski, 1930, pp. 65–7) there is something in this analysis. But the Maud analysis has an overly legalistic focus which obscures as much as it illuminates. Furthermore, if one also questions the 'top-down' view implicit in the Maud analysis (and also in Jennings's views) much of that analysis seems inaccurate. Let us examine some of Maud's assumptions.

Is legislation narrowly drafted? Most local government legislation enacted in the last fifty years, and particularly those Acts giving local authorities important social welfare functions, has been enabling legislation which imposes few specific duties or other legal constraints on local authorities. Authorities have been given *powers* rather than *duties* and many powers are subjectively formulated so that, subject to judicial review, they themselves define the limits of their powers. Certainly procedural restraints are imposed and central departments have substantial administrative powers, but in general local government legislation is not narrowly drafted. We have already seen that in relation to the management of council housing authorities have been given great discretion. Consider also the case of education. The basic legal framework is contained in the

Education Act 1944, under which the primary duty of the local authority is to ensure that 'efficient education . . . shall be available to meet the needs of the population of their area'. The nature of this legal framework has been commented upon by Professor Regan (1977, pp. 11–12):

> Although there have been a number of subsequent Education Acts, these have made relatively minor changes and the 1944 system remains fundamentally unaltered. A whole generation in England and Wales has been educated under its provisions. . . . Of course the structure of an education system does not entirely determine the content of the education provided. Since 1944 there have been profound changes in teaching methods, educational philosophy, school design, the role of head teachers, and much else. . . . Nevertheless these changes have occurred within an enduring framework.

Certainly the strains on this framework today are greater than ever but, given the changes in education policy, it is remarkable that the 1944 legislation has provided the basic statutory framework for the last forty years. Clearly this is not an example of narrowly drafted legislation.

Does legislation discourage enterprise and innovation? Since legislation generally provides considerable scope for discretionary action it is not obvious that enterprise and innovation is discouraged. Gyford (1984, pp. 8–9) gives several examples of local authorities using their powers in innovative ways – not merely by shifting expenditures but also by increasing the range and altering the nature of the services they provide. It could be argued that in recent years local authorities have enjoyed more room for manoeuvre than at the time of the case of *Attorney-General* v. *Fulham Corporation* in 1921, in which the attempt of the authority to transform their public wash-house into a municipal laundry was declared *ultra vires*. When local planning authorities felt that the judicial limitations placed on the use of conditions attached to planning permissions prohibited them from ensuring that developers contributed to the infrastructure costs of development, for example, they soon devised means, through the use of agreements, to circumvent these restrictions (Jowell, 1977; Loughlin, 1981). Also since 1963 local authorities have possessed various powers to enable them to act for the general good of their areas in ways not specified by statute (Crawford and Moore, 1983).

Furthermore, both Maud and Jennings seem to imply that policy originates from the centre. This is not necessarily the case. Indeed there is evidence to suggest that innovation originates at the local level: local authority discretion provides the opportunity for policy innovation; this leads to experimentation; this experience is recorded by the centre; eventually it is adopted by the centre and filtered back down to the localities through policy guidance or new legislation. This seems to have been the experience with comprehensive education, for example, where the early initiatives were taken by local authorities, often in the face of central ambivalence or hostility (Gosden, 1983, pp. 28–35).

Are local communities robbed of services? The *ultra vires* doctrine is undoubtedly more restrictive than some form of general competence power but the main effect, as Jennings points out, is not necessarily to hinder effective service provision but rather to require that in certain circumstances there be central approval for the new type of activity being promoted. In many cases the centre has not been unresponsive. Thus, when, in *Prescott* v. *Birmingham Corporation*, the Court of Appeal declared *ultra vires* the defendants' old age pensioners' travel concession scheme a Private Member's Bill, which became the Public Services Vehicle (Travel Concessions) Act 1955, was immediately rushed through Parliament. (Although compare the attitude of the Transport Secretary after the House of Lords' decision in *Bromley* v. *GLC* where he refused to introduce legislation validating the scheme and the supplementary precept.) On occasions very broad powers have been granted to local authorities through the promotion of private Acts: see, for example, the Zetland County Council Act 1974 giving the Shetland Islands Council unusual and extensive powers of control over the development of the Shetland Isles.

Is there an excessive concern with legalities? This argument has recently been dealt with authoritatively by Professor J. D. Stewart (1983, pp. 146–50). Stewart suggests that 'the myths of statutory constraint' have grown up to protect local services, organizational interests or key local actors and that the myth is often used in order to avoid the need to explore the nature of other constraints. Of course, under certain circumstances, such as the attempt by the current Labour administration at the GLC to promote a host of new initiatives, the issue of legality might be crucial, but in general this does not seem to be the case.

In general, therefore, there is little in the legal framework of local government which, within the functionalists' terms, necessarily prohibits the local authority from providing public services effectively. The crucial issue, as Jennings recognized, is that of the attitude of central government, since it has tremendous power to influence the behaviour of local authorities. For most of this period, and despite a persistent feeling that central departments have been overly involved in the details of local administration, there has been a basic consensus between departments and authorities over the need for public services and the manner in which they would be provided. In such an environment the 'top-down' views of the Maud Committee seem misplaced and relationships between central departments and local authorities are better characterized in terms of 'partnership', 'interdependence' and 'structured bargaining'.

During the last decade, however, this consensus, founded on the assumptions of growth, has been progressively breaking down. It is at this stage that it becomes clear that, whatever the merits of portraying central–local relationships in terms of bargaining, this focus cannot ignore the essentially hierarchical nature of the relationship. And it is during this period that the capacity of the centre to influence local authorities (a capacity which, as Alan Alexander explains in Chapter 3, has been increasing throughout the last fifty years) assumes new significance.

The result has been conflicts not merely over the level of services but also over the terms on which services are provided and indeed whether the local authority should be providing these services at all. The conflict became further intensified with the election of a Conservative government in 1979 which has tested to the limits the capacity of the centre to influence local authority action. Since 1979 there has been an unprecedented volume of local government legislation, which has affected virtually all areas of service provision. Even more important than the volume, however, has been the contentious nature of this legislation.

Basically the government has been attempting not only to reduce the level of local government expenditure but also to alter the nature of that expenditure. A primary objective has been to reform the grant system to incorporate incentives to reduce expenditure. But the government has also introduced various reforms which seek either to impose a surrogate market discipline on local public services or to encourage the privatization of local services. Examples include reforms to the housing subsidy system, which has been used by central government to provide incentives to local authorities to increase rents and make profits on the housing revenue account; the introduction of legislation requiring the competition with the private sector in such areas as bus routes or tendering for local authority projects; and the imposition of a planning framework on metropolitan authorities which requires them to justify in cost-benefit terms the amount of revenue support they give to public transport undertakings.

Consequently in recent years the consensual legal framework for local authority action has been breaking down. Local authorities still possess a great deal of power for discretionary action. But, by various strategic reforms, the government has sought to ensure that central government policies are implemented by local authorities. In some areas, such as council house sales or through the use of urban development corporations, this involved divesting local authorities of discretionary powers. But in relation to many local public services the existence of local discretion is an essential part of central government policy since, in the context of severe expenditure constraints, local authorities are presented with an invidious range of policy choices for which central government's responsibility is not always obvious.

Conclusions

The normativist and functionalist approaches to administrative law provide distinctive modes of analysing the tension between administration and law. The normativist theory of administrative law highlights the differences between administration and law and subordinates administration to law. Normativism focuses on the adjudicative role of law and therefore emphasizes its rule orientation and its conceptual nature. Fundamentally, it reflects the values of individualism and the ideal of the autonomy of law. On the other hand, the functionalist theory of administrative law all but collapses the dualism between administration and law. Functionalism does so by focusing on the regulatory role of law and consequently adopts a result orientation and an instrumentalist social policy approach.

Fundamentally, it reflects the values of collectivism and a theory of organic evolution.

In the period leading up to the publication of *Municipal Progress* administrative law thought was polarized between the normativism of Dicey and Hewart and the functionalism of Jennings, Laski and Robson. Subsequent legal developments affecting local government may be examined in terms of the relative influences of these two approaches.

In the courts the normativist orientation has been dominant. Up until the mid-1930s this was pursued in an activist manner and as a result an overly legalistic framework was imposed on local authority programmes, which hampered local action. After the mid-1930s, however, the courts appeared to recognize that regulatory programmes were a permanent feature of the modern state and consequently their adherence to a normativist approach required them to retreat from activism in judicial review. This period of quiescence lasted until the mid-1960s. Since then a new period of judicial activism has emerged, albeit within the shackles of normativism. (For an examination of the difficulties see Loughlin, 1978.) Finally, within the last decade, the incongruity of activism and normativism seems to have been recognized and, as a result, a functionalist approach is gaining support.

Within the pockets of administrative power left by the court-centered orientation of the normativists, functionalism has flourished. The challenge has been to establish an 'internal' administrative law. What resulted, however, as John Stewart explains in Chapter 5, was a technocratic conception of the ideal of good administration. The bureaucratic structures established were subject to radical functionalist criticism but in the 1980s reforms based on a normativist critique have been implemented. This normativist critique stems from a loss of faith in the capacity of local government to maintain 'continuous initiative' (Wallas, quoted by Young in Chapter 1) and thus challenges not only the technocratic conception of good administration but also the efficacy of public provision of services. Following the social market economy doctrines pioneered by Hayek (1960) and Friedman (1962) the Conservative government has sought to increase the sphere of society subject to the discipline of the market. For local government this means privatization, direct competition with the private sector on service provision, restraint on discretion through imposition of rules, individual rights vested in consumers of local services and a greater array of central control powers.

The degree of centralization of power resulting from these developments has been especially controversial. In seeking to impose this normativist framework on local government the Conservative government have taken drastic measures. In their relations with local authorities we have seen government by administrative diktat rather than in accordance with legal rules, gross manipulation of legal rules, retroactive alteration of legal rules and broad powers given to ministers in forms consciously designed to minimize the possibililty of judicial review. As a result there has been a breakdown in traditional practices; the idea of gradual and evolutionary change, notions of co-operation and co-ordination between the two tiers of government, the sense of restraint that comes from pluralism, and the acceptance of the limited efficacy of central direction. Indeed, there is a danger that in taking authoritarian measures, apparently to achieve freedom from the burden of the state the authoritarian state which is emerging might not easily be replaced by the normativist state. Thus, in Lord Hewart's terms we might be entering into a period of 'new despotism' (Hewart, 1929), although this trend might more accurately be examined in terms of Fraenkel's thesis on *The Dual State* (1941).

Administrative law thought therefore is currently within a period of crisis and change. The current trends towards functionalist judicial controls and normativist administrative controls are the very opposite of those operating in the mid-1930s. And the role and nature of governmental institutions is widely questioned. Some have responded by appealing to mysteriously pre-existent legal and constitutional principles (reformulated normativism: Dworkin, 1977; Elliott, 1981). Others adopt a nihilistic stance (extreme functionalism: Griffith, 1979). Nevertheless, an alternative to the emergence of an authoritarian state can be envisaged. This alternative, based on what may be termed a *proceduralist* theory of public law, would aim to establish both the structural premises for self-regulation *within* governmental bodies and the structure frameworks for relations *between* governmental bodies. This approach accepts that administrative processes are within the domain of law but rejects the instrumentalism and result-orientation of the functionalist approach and recognizes the normative character of law.

Administrative law would thus become a system for co-ordinating action between semi-autonomous governmental and quasi-governmental bodies. In the context of local government this approach builds on and transforms the radical functionalist critique of bureaucracy in local government and presupposes the breakdown of bureaucratic structures by such developments as tenant control of housing and community control of schools. It also implies a reorientation of central–local relations in the light of recent research highlighting the limited directive capacity of the centre (for example, Barrett and Hill, 1984). Obviously at the moment it is merely a vision, the details of which have yet to be thoroughly worked through (Teubner, 1983, 1984; Trubek, 1983). But at least it provides an alternative response to complexity and unmanageability of modern government structures than the centralizing and disengagement tendencies of current government policies.

7

Local Government Finance

RICHARD JACKMAN

> As a nation we have every reason to be proud not only of the progress of a century in municipal finance, but also of the actual stage of excellence which has been reached, for our fine system of local government finance will stand comparison with the systems in force in the local governments of any other country. Weakness of civic authority abroad may often be found to rest not a little in the faulty character of municipal finance, but in Great Britain we may rest satisfied that it is worthy of our best traditions of self-government, probity and ordered development.
>
> Sir Josiah Stamp in *A Century of Municipal Progress*, p. 399

As these closing sentences indicate, Sir Josiah Stamp's chapter on 'The Finance of Municipal Government' takes as its theme the achievements of the past rather than the problems confronting the system of local government finance in 1935. Those achievements had been remarkable. In the early nineteenth century, though the scale of local government activity was tiny, there were none the less numerous separate bodies each running a particular service and levying its own rate. By 1935 the immense growth in local government had been accompanied by a rationalization and consolidation of its finances. The various 'special-purpose' authorities, such as the School Boards, had been brought together into a single, all-purpose, local authority in each county or county borough and the rating system which provided for their finance had been restructured on a national and reasonably uniform basis.

Stamp portrays the contrast between the fragmented and rather primitive state of local government finance in the 1830s and the much larger and more organized machinery of the 1930s and it is in terms of this contrast that the achievements of the system are defined. When

Stamp refers to progress, his criteria are perhaps closest to those of the modern corporate planner, concerned primarily with the scope of operations and the efficiency and orderliness with which they are conducted. There is much more emphasis on the impact of local government finance on the internal operation of local government than on its external impact on the ratepayer or on the consumer of local government services.

Yet, in the 1930s as now, other types of questions *were* being asked about local government finance. Was local government spending, in aggregate, too high? Need central government take additional powers to control it? Was the provision of local government services equitable as between different households or different localities? Were local authorities providing value for money? What was the impact of industrial and commercial rates on the competitiveness of industry and on its location? Were rates on households excessively regressive? For reasons which are described by Young (Chapter 1), none of these issues receives even a mention, and it is no doubt the rather inward-looking approach that enables Stamp to reach so satisfied an assessment of 'the actual stage of excellence' attained by 1935.

During the fifty years that have followed the publication of *A Century of Municipal Progress* the administrative achievements of which Stamp was so proud have become increasingly taken for granted, while the effects of local government finance on the citizen and on the rest of the economy, which Stamp had largely ignored, have been, in one form or another, a continuing and increasingly serious source of discontent. Stamp's commendation of the achievements of 1935 stands in sharp contrast to the much-criticized system of local government finance today but the reason is not because the system has deteriorated but that the criteria for evaluating it are different and more exacting.

The task of this chapter is to document the main changes in local government finance over the last fifty years. A major aim is to attempt to understand why we have still not been able to establish in this country anything approaching the 'fine system of local government finance [which] will stand comparison with the systems in force in the local governments of any other country' which Stamp thought had already been attained in 1935. That failure has at root been a failure of central government. Successive central administrations have preferred to keep as many levers of political control as possible in their own hands rather than create the financial environment within which autonomous and politically independent local authorities could flourish. In particular, central government has used local authorities as agents of its national policies for income redistribution, operating through various agencies of the welfare state. The result has been to break the link between the benefits of local government spending, which are received to a large extent by some sectors of the community, and the costs which are largely borne by others. In the older towns and cities, Labour politicians have sought to expand services for the poor or those in need knowing the cost will be borne by others, in practice largely by local industry or central government grant. The consequence has been to undermine the financial accountability of urban authorities. Initially, the rapid growth in expenditure was supported by central government grant but by the late 1970s the taxpayer's willingness to pay had become exhausted and central government policy was thrown into reverse. While lack of financial accountability in local government has been the main cause of high spending, and hence the cause of the present difficulties, the policies which Conservative administrations have been pursuing since 1979, far from attempting to revive local accountability, have increased central intervention and central control. From a different perspective, one could argue that some local authorities had become so addicted to grant, and were finding such difficulty in coping with grant withdrawal that they have had to be taken into government care.

To explore these developments, this chapter starts, as did Stamp's chapter of 1935, with changes in local government spending over the period. It documents both the growth in expenditure, and the replacement of the trading services of the 1930s by the social services of today. The rating system, too, has become more progressive, as a result of the introduction of rate rebates. Taking expenditure and rates together, local government has become – as it was not in the 1930s – a major instrument of income redistribution. Grants have increased very substantially as a source of local finance, but the grant system has been designed to achieve other objectives rather than to enhance local accountability. The policies of the present government are concerned more with its national policy objective of reducing public expenditure than with remedying the structural weaknesses of local government finance which have led to the present plight.

Table 7.1 *Local government spending and GNP, Great Britain (selected years)*

	Local government spending % of GNP			Real local government current expenditure per head (1975 prices)
	Total	Current	Capital	£
1890	4·5	3·9	0·6	29
1910	8·6	6·6	2·0	57
1930	13·7	11·0	2·7	95
1935	13·2	11·2	2·0	110
1940	12·1	10·1	2·0	(111)[1]
1945	9·7	9·3	0·3	(88)[1]
1950	11·3	8·1	3·2	82
1955	11·7	8.1	3·5	95
1960	12·0	9·2	2·8	116
1965	14·7	10·3	4·4	148
1970	18·2	13·7	4·5	217
1975	18·4	13·7	4·8	241
1980	17·2	14·4	2·8	265
1982	18·5	16·1	2·4	290

Note[1]: no price indices were computed during the Second World War. The figures in column (4) thus relate to 1938 and 1946 rather than to 1940 and 1945
Sources: Local government spending 1890–1975, Foster *et al.* (1980) pp. 102–6; 1980 and 1982 *Annual Abstract of Statistics*. GNP, prices, population up to 1945 from Peacock and Wiseman (1967); subsequent years from *Annual Abstract of Statistics*

The Scale of Local Government Expenditure

At the close of the century of expansion which Stamp had described, 'Municipal receipts per pound of income per head increased eight and a half times, and burden of rates per pound of income per head increased ten times. These are, of course, only broad computations designed to indicate the immense difference in the scale of municipal service and demand' (Stamp, in *Municipal Progress*, p. 372). By 1935 local government expenditure amounted to just over 13 per cent of gross national product (GNP). By 1982 that share had increased to about 18·5 per cent. These figures appear to suggest that, while local government is still growing, the momentum of expansion has slowed down and the rapid growth of the earlier years has given way to a more sedate rate of growth since the 1930s. But such an impression is misleading: local government has been subject to a number of major changes over the last fifty years.

The main trends in aggregate expenditure are summarized in Table 7.1. The table shows gross local government expenditure as a proportion of GNP and real per capita gross current expenditure. It should be stressed that the gross measure of local government spending is inclusive of services financed by fees and charges; by contrast the net, or public expenditure, definition of local government spending excludes such services. Thus gross expenditure gives a better picture of the overall scale of local government activity, while net expenditure measures its cost to the ratepayer or taxpayer.

During the Second World War there was, of course, a fall in local government spending, primarily on capital, but this fall was quite quickly recovered and by 1947/8 local government spending was back to its prewar levels. Of greater long-term consequence for local government were the reforms of the postwar Labour government which took away from local government a variety of its traditional functions, most of which were centralized in the hands either of central government itself or of public corporations acting at a national level. In this process local government lost services (described in more detail in the next section) which had amounted to about one-quarter of its total expenditure before the war.

The twenty or so years that followed, from the early 1950s to the early 1970s, seem in retrospect a time of steady and untroubled expansion. A number of the services left with local government, in particular education and housing, turned out to be amongst the most important and rapidly growing elements of the postwar welfare state. Real local government spending trebled between 1950 and 1975, equivalent to an annual average growth rate of 4·7 per cent. By comparison, the tenfold increase in local government activity between 1835 and 1935 is equivalent to an annual average growth rate of 2·3 per cent. By 1975 local government spending amounted to over 18 per cent of gross domestic product (GDP), even though in 1974 local government had lost two further services, water and sewerage and local health services.

The era of expansion was brought to an end primarily by external events: the increase in oil prices introduced by OPEC in the autumn of 1973 and the high rates of inflation and unemployment which followed it led national governments in the UK (as elsewhere) to adopt much more restrictive policies on public expenditure, including local government expenditure. But if the initial check to expansion came from external circumstances, it was to be followed, with the advent of the Conservative government of 1979, by an attack on local government spending based more on ideology than economic necessity. The Conservatives wanted, in terms of ideological principle, to cut back the welfare state. Local authorities, which previous governments had made the mechanism for the growth of the welfare state, were now expected to go into reverse and to cut back the services they had previously been required to build up.

The late 1970s thus saw a fall in local government spending, but closer inspection (Table 7.1) shows that that fall was concentrated almost entirely in capital expenditure while current spending continued to grow. The more intense pressure since 1980 to cut expenditure appears to have had the opposite effect. The growth of local government spending has slowed down, but the severity of the recession since 1980 has led to a much sharper fall in GNP, so that total gross expenditure as a proportion of GNP has risen.

Local Services

Looking back over the past century, William Robson observed that 'the energies of local authorities were directed primarily to the conquest of the material environment. Drains, sewers, infectious disease, public cleansing, gas, water supply, electricity, tramways, omnibuses, highways, parks, fire brigades – these are the characteristic items on the agenda papers of the nineteenth-century councillor' (*Municipal Progress*, p. 459). Robson's picture of the Victorian local authority is not strictly accurate. The most ancient of local authority services, poor relief and the police, were major items on local authority budgets throughout the nineteenth century. During the period of great expansion of local government from 1870 to 1900, the most important new service to be taken on was not gas or water, but education. Even so, there is no doubt that the most conspicuous achievement of local government in the hundred years up to 1935 lay in the improvement of the material environment of the towns and cities.

In 1935 the services on Robson's list absorbed about half the gross current expenditure of a typical local authority. The proportion appears to be greater than at any time during the nineteenth century. Local authorities were continuing to increase their expenditure on these 'material' services throughout the 1930s, the most rapid growth at that time being in outlays on local hospitals and electricity. The main new item on the agenda paper of the 1935 councillor was housing. Before the First World War, local government expenditure on housing had been negligible, but after the war, local authorities were made the agents of Lloyd George's 'homes for heroes' housebuilding campaign. By 1935 housing accounted for close to 10 per cent of current expenditure and for over a third of the total capital expenditure of local authorities in England and Wales.

Thus by 1935 the main thrust of local authority activity continued to be, in Robson's phrase, 'the conquest of the material environment'. Traditional functions such as police and poor relief remained (though the latter now in uneasy relationship with the national unemployment insurance and pension schemes). But already by 1935 local government was substantially involved in two services not on Robson's list and which were to become central elements of the new welfare state, namely education and housing.

Fifty years later, the transformation of local government activity is immense. The changes, in terms of gross expenditure, are set out in Tables 7.2 and 7.3. Most of the services in Robson's list have disappeared. Electricity and gas were taken over by public corporations in 1947 and 1948, respectively. Local hospitals were taken over by the national health service in 1947, as were other local health services in 1974. Central government assumed responsibility for trunk roads in 1936 and 1946. Water supply and sewerage were transferred to regional water authorities in 1974. By 1982 the services on Robson's list accounted for little more than 10 per cent of local authority current expenditure.

In their place have come many of the major services of the postwar welfare state. Education, propelled forward by the 1944 Act, has become by far the largest local authority service in terms of current expenditure. Real per capita current expenditure on education in 1982 was five times what it had been in 1935. The thirty years after the Second World War saw an unprecedented rate of local authority housebuilding. By the late 1970s slum housing had been virtually eradicated and housing conditions – at least in terms of physical amenities – transformed. Towards the end of the period, during the 1970s, local authorities had also embarked on a major expansion of social services.

Clearly then there has been a great change in the types of activities in which local authorities have been engaged over the last fifty years, which can perhaps be crudely characterized as a shift away from the provision of trading services ('gas and water socialism') towards the

Table 7.2 *Local government current expenditure 1930–82, major services (% of total current expenditure)*

Service	Year											
	1930	1935	1940	1945	1950	1955	1960	1965	1970	1975	1980	1982
Education	19·8	19·1	17·0	17·6	30·3	33·4	37·3	37·9	35·1	35·9	36·1	36·1
Highways and lighting	13·0	11·5	8·3	5·8	8·4	7·9	7·0	6·9	5·3	4·0	6·8	7·0
Housing	8·3	9·8	8·2	6·6	9·5	13·4	15·0	15·0	14·7	16·6	21·8	17·2
Poor relief	8·0	8·0	6·1	4·1	—	—	—	—	—	—	—	—
Police	5·1	4·9	5·4	4·8	5·7	5·9	5·8	6·1	5·6	5·4	6·5	7·0
Hospitals	2·7	5·0	5·7	5·9	—	—	—	—	—	—	—	—
Sewerage	2·4	2·5	2·2	1·9	2·1	1·9	2·1	2·3	2·4	—	—	—
Refuse	1·7	1·6	1·5	1·8	2·4	2·3	2·0	1·9	1·8	1·9	2·1	2·0
Parks	1·2	1·3	1·2	1·0	1·6	1·5	1·5	1·3	1·2	1·3	3·0	2·5
Fire	0·5	0·6	0·6	0·7	1·6	1·6	1·5	1·5	1·2	1·3	1·5	1·5
Libraries and museums	0·5	0·6	0·6	0·6	1·0	1·0	1·1	1·1	1·1	1·1	1·2	1·1
Local health services	—	—	—	—	3·8	3·7	3·7	3·6	3·1	—	—	—
Welfare (social services)	—	—	—	—	2·9	3·3	3·1	3·2	3·0	6·1	6·9	7·1
Trading services	25·6	25·6	25·9	30·2	15·8	13·3	11·3	9·5	7·2	2·4	2·5	2·6
of which:												
Water	4·5	4·5	4·2	3·9	4·2	3·7	3·4	3·0	2·9	—	—	—
Gas	4·0	3·5	3·2	4·2	0·4	—	—	—	—	—	—	—
Electricity	6·8	8·9	10·3	12·0	—	—	—	—	—	—	—	—
Public transport	6·3	4·8	4·3	5·0	5·9	5·1	4·1	3·1	2·1	0.6	0·5	0·5
Harbours	3·1	2·8	2·5	2·9	3·0	2·7	2·4	2·0	1·3	0·2	0·2	0·2

Sources: 1930–75 Foster *et al.* (1980) pp. 102–15; 1980 and 1982 *Annual Abstract of Statistics*

Table 7.3 *Local government capital expenditure 1930–82, major services (% of total capital expenditure)*

Service	Year											
	1930	1935	1940	1945	1950	1955	1960	1965	1970	1975	1980	1982
Education	5·5	5·3	12·3	1·7	8·5	11·4	5·1	11·0	12·9	10·2	7·6	8·8
Highways and lighting	14·9	9·1	8·6	5·9	1·6	2·0	5·4	6·4	9·9	7·6	9·5	12·4
Housing	39·3	36·3	24·9	32·9	70·9	68·5	56·9	59·0	48·4	65·9	63·2	54·0
Poor relief	0·8	1·0	0·9	0·4	—	—	—	—	—	—	—	—
Police	0·4	0·3	0·1	0·4	0·7	1·2	0·6	0·7	1·0	0·9	0·9	1·2
Hospitals	2·7	1·5	5·7	0·8	—	—	—	—	—	—	—	—
Sewerage	4·1	6·8	4·9	1·7	2·9	3·7	6·2	5·1	6·7	—	—	—
Refuse	0·3	0·3	0·3	—	0·2	0·1	0·2	0·2	0·7	0·5	0·8	1·1
Parks	1·6	1·7	1·2	0·4	0·4	0·4	0·5	0·8	0·7	1·8	2·4	2·9
Fire	0·2	0·5	0·6	—	0·1	0·3	0·4	0·3	0·3	0·2	0·2	0·4
Libraries and museums	0·2	0·4	0·3	—	0·1	0·1	0·3	0·3	0·4	0·5	0·4	0·4
Local health services	—	—	—	—	0·5	0·3	0·4	0·7	0·8	—	—	—
Welfare (social services)	—	—	—	—	0·9	0·6	1·2	1·3	1·2	1·8	1·6	2·0
Trading services	25·7	27·5	25.0	44·1	9·3	6·9	8·0	6·5	8·9	2·9	3·5	5·2
of which:												
Water	4·7	5·0	5·4	5·0	4·6	4·5	4·9	3·4	4·1	—	—	—
Gas	1·4	1·5	1·5	3·8	0·2	—	—	—	—	—	—	—
Electricity	14·0	15·7	12·7	28·2	—	—	—	—	—	—	—	—
Public transport	2·2	2·2	1·8	2·9	2·7	0·6	0·4	0·2	0·6	0·1	0·3	0·3
Harbours	1·6	1·2	1·3	2·5	1·0	1·0	1·3	0·8	0·7	0·3	0·2	0·3

Sources: 1930–75 Foster *et al.* (1985) pp. 102–15; 1980 and 1982 *Annual Abstract of Statistics*

array of services of the postwar ('cradle to grave') welfare state. This change was not the outcome of any conscious process of evolution within the local authorities themselves, nor can it be explained as an efficient functional division of responsibilities between central and local government. Rather, the new pattern was imposed, sometimes on rather arbitrary grounds, by the postwar Labour government of 1945–51. There is no obvious logical reason why, for example, health should be a national service while education remained local, nor why gas should become national while water remained local. It is easy to get the impression that Attlee's government would have liked to take over all the major services, and only the immensity of the administrative costs dissuaded it from doing so. It therefore took over the services in which it had most interest, leaving the rest to local government.

The outcome, in any event, was to take away from local government its most obviously 'commercial' services. While some of the services left with local government, in particular housing, could have been operated as trading services, it was never intended that they should be. Many of the services local authorities now provide benefit not the community as a whole but specific groups within it. Local government services have thus become to an increasing extent redistributive, and it is the consequences of this change which, it will be argued, have underlain many of the difficulties which have arisen in local government finance over the past fifty years.

Rates

Introducing his review of the sources of local finance, Stamp remarks that:

> We have to consider the several sources of revenue and the chief and the oldest (apart from fees and property) is the rate levied on property. No space need be given to the description of the origin or nature of an institution or a method so common and so well understood. (*Municipal Progress*, pp. 378–9)

It is understandable that Stamp should have thought it unnecessary to discourse at any length on the origins of the rating system. The second edition of Edwin Cannan's classic study *The History of Local*

Rates in England had been published in 1912 and would have been regarded in the 1930s (as it still is today) as the final word on that subject. The nature of the rating system is quite a different matter. The late 1920s had seen the most radical upheaval in the history of British rating culminating in the derating of industry and agriculture in 1929. Stamp devotes only three and half pages out of a thirty-page chapter to the rating system, and most of that is on valuation (see below).

During the 1920s the rating system had come under attack primarily because of the belief that it was exacerbating the depression in industry. When industry went into decline, local government expenditure on poor relief for the unemployed would rise, while its tax base (local rateable value) would tend to fall. To balance the budget, rate poundages would have to rise, throwing a heavy rate burden on to the surviving firms. This was held to constitute a serious handicap for British firms relative to their foreign competitors, and hence to have intensified the depression in industry.

To make matters worse, industrial decline was concentrated in particular localities ('necessitous areas') so that rate poundages in those areas tended to be especially high. As a consequence of high rates, new investment was driven out of the areas which needed it most, a process graphically described by Winston Churchill (then Chancellor of the Exchequer) in 1928:

> What with the ever growing burden of the rates and the poisonous atmosphere created, industry flies from these districts or dies within them. Who would start a factory in West Ham? On the contrary everywhere we see industries shifting outside heavily rated areas. . . . Frequently we see the new factory erected just outside the municipal boundaries to escape from being made a starveling milch cow. The whole impulse is for industry to quit the melancholy areas and leave behind them political agitation, State assistance and a forlorn mass of derelict humanity. (cited in Mair, 1983)

There were a number of ways of dealing with these problems. Central government could take over the provision of poor relief, thus removing a major burden from the worst hit areas. Central grants to local authorities could be increased to give more help to the most needy authorities. While both these reforms were to be introduced in stages over the next fifty years, in 1928 Churchill chose a more direct and radical solution. Industry was 75 per cent de-rated and agriculture, which also suffered severely in the depression and was already 75 per cent de-rated, was exempted from rates completely. While the main purpose of these measures was to assist industry, de-rating also had important implications for local government finance. It meant that the bulk of rate revenue was collected from households. While there appear to be no exact statistics, according to Hicks *et al.* (1944, p. 3) during the 1930s about two-thirds of total rate receipts derived from households.

Further, the incidence of the rate burden amongst households was well known to be regressive. The Liberal Industrial Inquiry of 1928 had found evidence that a working man on a low income might be paying as much as 10 per cent of his income in rates, while for a rich man the figure was close to 1 per cent. In their much more ambitious and systematic study of the situation in 1937, John and Ursula Hicks found that, amongst working-class households, rates amounted on average to nearly 5 per cent of income for households at the lower end of the wage scale but to less then 3 per cent of income to those towards the upper end (Hicks and Hicks, 1945, pp. 24–5). However, their study covers only working households and thus leaves out both the poorest households (the elderly and the long-term unemployed) where rate burdens are the highest and also the wealthiest households with the lowest rate bills in proportion to income.

It is none the less clear that the burden of rates on working-class households and on others on low income was much higher relative to income than it is today, when the average for low income households is of the order of 2 per cent. This fact must of itself have been a powerful spur to careful financial management and economy in local government spending. Indeed, it was argued that the objectives of autonomy and financial accountability in local government were best served by a regressive tax system. Thus Hicks *et al.* (1944, p. 10) defended the regressiveness of the prewar rating system from the perspective of a poor local authority:

> At present, if its citizens decide that there is some amenity which they desire badly enough for them to be willing to meet its cost out of their limited incomes, then they can have it, each of them bearing his share in the cost. But with a local income tax, the greater part of the cost would have to be borne by a few large or

> moderately large incomes. . . . Even if the revenue was obtainable on paper, it would usually be obtainable only by imposing a high rate of tax – and a high rate of income tax would be vastly more effective than high rates in driving people out of the district.

It is a well-known principle of local public finance that local authorities cannot satisfactorily carry out independent policies of income redistribution. If one local authority is more generous than others the poor tend to move in to take advantage of the better provision while the wealthy tend to move out to avoid the higher tax burdens so that, ultimately, it becomes impossible for the locality to afford to continue with the more generous policy (Oates, 1972, ch. 1). This problem had been recognized from early times: since the sixteenth century, legislation had allowed local justices to move on the vagrant poor and the Poor Law Act 1834 attempted to standardize the provision of poor relief through the creation of 'unions' of parishes (Foster *et al.*, 1980, pp. 51–2).

But Hicks's argument goes further. Even if local government services are not of themselves redistributive, financing them through a progressive tax makes them so. An increase in local government spending financed by progressive taxation automatically entails an increased degree of income redistribution. If local authorities are alert to this problem, their freedom to determine their spending levels is effectively curtailed. If they are not, the outcome is an increased degree of polarization as the poor congregate in the high-spending local authorities and the rich move out to the low-spending ones.

These warnings notwithstanding, the main developments in the rating system have been designed to protect and conceal from the domestic (household) ratepayers the ever-increasing cost of local government services. One major element in this process has been the growth in government grants (see below). Within the rating system itself the main changes have been the re-rating of industry (but not agriculture) in 1958, the rate rebate scheme introduced in 1966 and the 'domestic element' of the rate support grant, essentially a form of 'super-rating' of industry and commerce, introduced in 1967. The impact of the first and third of these measures was to reduce substantially the proportion of rates paid by households – from about two-thirds before the war to about 44 per cent today (Department of the Environment, 1981, para. 1.2).

More fundamental, however, has been the impact of rate rebates, the result of which has been that many low-income households are relieved of the whole (or part) of their rate bills. By 1981 nearly 6 million households (about 30 per cent of all households) were in receipt of rate rebates or supplementary benefit rate relief (ibid., para. 4.3). Rate rebates depend on family and other circumstances and on the level of rates, so that not all on low incomes qualify for them. None the less, their effect has been to end the regressiveness of the rating system at the lower end of the income scale. In many local authorities, a substantial minority of the population is now wholly (or largely) exempt from rates. Further, this same group of households often receives a disproportionate share of local government expenditure (for example on housing or social services).

Thus there has been a significant change in the nature of local government activity over the last fifty years. While in the 1930s its activities were largely collective, providing services for all its citizens for which all would pay, they are now more redistributive. Often those for whom most services are provided pay little or nothing in rates, while those bearing the heaviest rate burdens often receive virtually no benefit from local government services. Hicks *et al.* (1944a, p. 11) had suggested that, in a system in which local authorities were enabled to make redistributive choices, 'a very disagreeable element of division would be introduced into local politics. . . . The economic conflict between classes is bound to cause trouble in a society such as ours; but it will cause far less disunity if it is played out on the national stage'.

By the late 1970s domestic rates were financing less than 20 per cent of the cost of local government spending. In urban authorities – and indeed elsewhere – local government spending could be increased in the awareness that the bulk of the cost would be borne not by local residents but by industry, commerce and central government. While expenditure grew most rapidly in the Labour-controlled urban authorities, the redistributionary potential of the local government system did not appear to be explicitly recognized by politicians in central or local government. The recent claim by the leader of the Greater London Council, Ken Livingstone, that the 'rating system is the best method of redistributing wealth that the labour movement has ever had its hands on' (Boddy and Fudge, 1984, p. 265) and the policies which the GLC has recently pursued of increasing expenditure for the explicit purpose of redistribution are perhaps the most overt recognitions of how the present local government finance system operates.

While rate bills generally form a relatively small proportion of business costs, and can be accommodated easily in times of business prosperity, in the recent recession industry and commerce have become much more concerned about rate bills, and particularly about the high rate poundages in some high-spending Labour authorities. Thus we almost have a repetition of the situation of the 1920s in which industrial decline is exacerbated by the redistributive policies and consequent high rate poundages of Labour authorities in the least prosperous localities. Unlike Churchill in the 1920s, the present government has chosen not to assist industry by relieving it of part of its rate bills. Instead it has tried to cut local government spending by cutting central government grant, particularly to the urban authorities, in the process raising rate poundages and thereby increasing rather than relieving the burden on industry.

Valuation

One of the most difficult and contentious aspects of the rating system has been the problem of valuation: the procedure for assigning to properties the rateable values upon which rate liabilities are based. The legal definition of rateable value, the net annual rental value of the property, had been clearly established for many years (at least since the Parochial Assessments Act 1836) but in practice rateable values have always tended to fall a long way below this level. In the nineteenth century there had been three main reasons for this shortfall. First, valuation was administered locally with no uniformity of procedure; second, revaluations were irregular and infrequent; and, third, various deductions and exemptions were allowed which varied from one locality to another. This last problem was exacerbated by the fact that in the nineteenth century a large number of different authorities were empowered to levy rates on a particular property for various purposes, and these authorities would often allow different deductions.

Towards the end of the nineteenth century, a major impetus to improve the consistency and accuracy of rating valuations came from the Inland Revenue, which itself had to make assessments of the rental value of property for Schedule A of the income tax. While local authorities were not obliged to follow the Inland Revenue valuations they were under pressure to do so to ensure that the various parishes within a county each made a fair contribution to county council services. This administrative advance was accompanied by pressure for legislative reforms which, though to a large extent inspired by Goschen in the 1870s, took practical effect only in the Rating and Valuation Act 1925, which provided for regular quinquennial revaluations, a national uniform schedule of deductions and the consolidation of rating authorities.

These measures in principle resolved the main problems of the nineteenth century valuation system and led Stamp to claim: 'The process of raising the valuations to the full rental values has proceeded more or less steadily through the last half century, and the technical perfection of the valuation has been mostly completed in the last two decades' (*Municipal Progress*, p. 379). Much progress had been made, which Stamp describes in his chapter (pp. 379–81), yet new difficulties were soon to arise, and to believe the system was approaching a state of 'technical perfection' was something of a delusion. The attempt at a general revaluation in 1928 had revealed a major new problem which has bedevilled valuation ever since. As a result of the combination of the rapid rise in prices during the First World War and rent controls on the bulk of the housing stock introduced at the same time, the actual (controlled) rents on most houses had fallen significantly below the notional free market rents upon which rateable values were to be based. Valuers faced with a dearth of eligible rental evidence fell back on a variety of expedients, such as using prewar rents.

The system of rating valuation in the 1930s, far from being close to perfection, was, as the carefully documented study of Hicks *et al.* (1944) shows, riddled with anomalies and distortions. New houses tended to be more lightly assessed (to the benefit of the ratepayer) than old. All property in wealthy areas tended to be more lightly assessed than property in poor areas. As Hicks is driven to conclude: 'The fact is that there is absolutely no guarantee that similar pieces of property will have similar rateable values, nor even that what is obviously a better house will have a higher rateable value than a house of lower quality' (ibid., p. 2). Hicks recomended that responsibility for valuation be transferred to a central authority (ibid., p. 72), and in 1950 valuation became the responsibility of the Inland Revenue Valuation Office.

The principle of quinquennial revaluations laid down in the 1925 Act was not to become a reality. For reasons of administrative costs, or political convenience, successive revaluations were postponed or cancelled. The revaluation due in 1933 was abandoned to save public expenditure, that due in 1938 was overtaken by the war and not completed until 1956. The two subsequent revaluations, of 1963 and 1973, were the only comprehensive attempts to value the national property stock at current notional market rental values, as required by the 1925 Act.

But by 1973, following the Rent Act 1968, a free market in rental housing had all but disappeared. The valuation system had lost any basis in reality, in the form of market rents, and had instead become based on various criteria, of different degrees of arbitrariness, for comparing one property with another. In 1976 the Layfield Committee (p. 169) reported:

> The Valuation Office of Inland Revenue advice is that in their belief the point has now been reached in England and Wales where enough rental evidence to support another revaluation on the present basis will not be available. This view is supported by all the professional bodies of valuers concerned with rating.

In the absence of any information on market rents it is of course impossible now to make direct comparisons between rateable values and market rents as Hicks had done for the 1930s. But it is possible to compare rateable values with capital values of properties (Evans, 1976; Foster *et al.*, 1980, pp. 309–37). Such studies show that rateable value relative to capital value is low (to the benefit of the ratepayer) for the most valuable houses and, for all types of houses, in rural as against urban localities. While these findings might in part be explained by the fact that the relationship of capital to market rental value will not be identical for all classes of property, there must, at the least, be a strong presumption that the anomalies in valuation of the 1930s have yet to be set right. In the absence of adequate evidence, the Inland Revenue valuers have to a large extent followed traditional valuations and thereby preserved the inequities of the past.

Since 1973 matters have gone from bad to worse. The traditional rating system lies, as Prest (1983) has put it, 'in ruins'. The Layfield Committee in 1976 had recommended that future revaluations, at least of domestic property, be carried out on the basis of capital values. The then Labour government equivocated, not least, perhaps, because it did not have a parliamentary majority with which to carry through legislative reforms. The 1979 Conservative government postponed the revaluation due in 1983 on the pretext of the party's pledge to abolish – or at least to reform fundamentally – the domestic rating system. The official announcement that the government would renege on that pledge was not made until the summer of 1983, and, according to the Rates White Paper (Department of the Environment, 1983), the government were at that time 'considering urgently the case for a domestic revaluation and will issue a consultation document as soon as possible' (para. 5.10). At the time of writing, twelve months later, nothing further has been heard of this urgent consideration nor of the consultation document. The government's dilemma is evident: a domestic revaluation based on rental values is an evident nonsense but one based on capital values would shift the burden of rates on to the wealthy and on to those living in the country; that is on to Conservative Party supporters.

None the less, the failure to do anything about valuation, together with the failure to reintroduce rates on agriculture, has meant that, in the recession of the 1980s, the burden of rates has fallen most heavily on the urban and industrial areas of the country – that is, on those least able to afford it – while the more prosperous areas of the country – typically outside the older towns – are especially favoured.

Grants

> Grants-in-aid have developed from the simplest beginnings a century ago into one of the most powerful engines of administrative machinery, and are today very much more than mere financial devices for spreading the burden over the wider field than the local ratepayers. (Stamp in *Municipal Progress*, p. 382)

The history of grants in the century up to 1929 had been one of an almost continuous proliferation of specific grants linked to particular services checked only by the short-lived Goschen reforms of 1887. By the early years of the twentieth century there existed an immense variety of different grants each administered according to different rules and the system as a whole had become so obscure and complex as to be virtually incomprehensible. In 1920, Sidney Webb, in the introduction to his book on the grant system, was driven to conclude that 'How exactly the grants are allocated among the different local authorities seems past all finding out' (Webb, 1920, p. 7).

The 1929 Act represented a major effort at rationalization. A large number of specific grants were scrapped and in their place was instituted a 'block grant', that is, a grant to a local authority of a specific sum of money which is not dependent on its expenditure on (or provision of) any particular service. The 1929 block grant took the form of a per capita grant weighted according to the local authority's unemployment rate, proportion of children under 5 years old, road mileage and shortfall of rateable value below a nationally determined standard.

The 1929 Act thus introduced three principles of grant distribution which have become central to the development of the grant system up to the present day. The first is the principle of a block, or 'unhypothecated', grant which carries with it the implication that it is for the local authority to decide how the grant money be spent in accordance with its perception of the relative needs for different services within its area. The second is the principle of basing the grant on an assessment of the local authority's spending needs, which in 1929 were taken to be children under 5, and so on. The third principle is that of assisting local authorities with low rateable value per head.

Despite these major innovations, the grant system introduced in 1929 had little practical impact on the distribution of grant, for two main reasons. First, while a large number of small grants were consolidated into the block grant, a small number of large grants, particularly those for education, police, highways and housing, remained outside the block grant system. These specific grants which remained absorbed about 75 per cent of the total grant (Carlson, 1947, p. 32). Second, the allowance for low rateable value in the formula turned out to be rather small, so the assistance provided to authorities poor in rateable value was correspondingly meagre. In an early study, Newcomer (1937) found that the variation of rate poundages after the 1929 Act was as great as it had been before.

Successive reforms of the grant system over the past fifty years have, however, been based on the framework introduced in 1929. It was not until 1960, when the grant for education was absorbed, that the block grant has become financially dominant (since when it has accounted for over two-thirds of the total grant distributed). The assessment of local spending needs has become increasingly detailed, until today when hundreds of factors are taken into account. Further, successive reforms of the grant system between 1947 and 1980 have effectively replaced the idea of helping poor authorities (in terms of rateable value) by a complete equalization, so that now all authorities (except the very wealthiest) are fully compensated for any shortfall (or surplus) of rateable value they may have relative to a given national standard. Since 1974 the grant system has been built explicitly around the concept of equalization: that is to say the principle of compensating local authorities in full for differences in their spending needs and in their rateable values.

Very recently, there has been a significant step away from equalization, in the form of the expenditure 'targets', and associated grant penalties introduced in 1981. The purpose of this device is to use the grant system as a financial pressure to reduce spending, rather than as a means of equalizing tax burdens.

The various reforms to the grant system between 1929 and 1980 have brought about an impressive progress in terms of logic and coherence. Yet, despite these achievements, the grant system of the 1980s has been a source of great dissatisfaction to both local and central government, and a major cause of conflict between them. While one main reason has been the attempt by central government to use the grant system to reduce local government spending, the progress towards equalization has also exposed structural weaknesses.

In practice, the concept of equalization has meant the equalization of rate poundages across authorities providing more or less equivalent standards of services ('equal poundages for equal services'). But the more successfully rate poundages are equalized the closer becomes the relationship between rate bills and rateable values. Anomalies and inconsistencies in the assessment of rateable values will generate under a system of rate poundage equalization corresponding inequities in rate bills. Thus the move towards equalization in the grant system has sharpened the tendency of the valuation system to favour prosperous and growing areas of the country at the expense of areas of economic difficulty. The towns and cities which were at the forefront of economic progress in the 1930s are today in decline and further disadvantaged by the combination of obsolete valuations and 'equalizing' grants.

The concept of spending need, upon which grant payments are based (though the concept itself has been renamed 'grant related expenditure' in the post-1980 system) has become increasingly political. As the services provided by local authorities have become increasingly redistributive, the perception of the 'need' for a service may differ rather substantially between those who receive the service and those who have to pay for it. These differences have been taken up by the political parties with Labour, representing the recipients of local government services, becoming the party of high spending in local government and the Conservatives, representing the ratepayers, the party of low spending. The gulf in expenditure per head between Labour and Conservative authorities has become very large. A Labour central government between 1974 and 1979 was prepared to support the high spending of Labour local authorities with grant, because it shared their perception of their high spending need. Since 1979, the Conservative central government, with a different perception of spending need, has taken away grant from the high spending, mostly urban, Labour authorities. In so doing it has provoked a major political confrontation between central and local government. The Labour local authorities see their grant as wholly inadequate in relation to their perception of their spending needs, while central government regards their expenditure as excessive in relation to its perception of their needs.

The government has, however, wanted to achieve a substantial reduction in local government spending overall, and not just in the Labour authorities, to reduce public expenditure and the size of the public sector in the economy as a whole. This determination from the centre to control local government expenditure is perhaps the greatest contrast with the position of the 1930s, as described by Young (Chapter 1). Initially it has tried to make use of the grant, by superimposing on to the block grant introduced in the Local Government, Planning and Land Act 1980 a system of expenditure targets and financial penalties to deter high levels of local government spending. The cut-backs in grant have, however, led most authorities to raise their rates in order to maintain their spending, rather than to cut their spending as the government would have wished. Whether this outcome should be attributed to the unresponsiveness ('irresponsibility') of local authorities to financial pressures, as the government would claim, or to technical deficiencies in the design of the grant system remains open to doubt (Gibson, 1982). But the failure of financial penalties operating through the grant system to reduce local spending has led in 1984 to the Rates Act, which for the first time empowers central government to set a maximum rate that can be levied by individual local authorities.

It would be inappropriate to discuss the technical details of the Rates Act here (Jackman, 1984). From a strategic viewpoint, the key issue is whether the Rates Act can be envisaged as a temporary 'holding operation' to control local government spending from outside while a sounder and more effective internal financial structure is being put in place, or whether it represents a permanent shift in the balance of power between central and local government. There are no indications that the present government has any plans for radical reform of local government finance consistent with increasing local accountability or tolerating greater local autonomy. While the present preoccupation with reducing overall public expenditure at the expense of all other policy objectives will no doubt diminish, the prospects for any central government initiatives to revitalize local government none the less appear very dim.

It is an interesting footnote to a discussion of the balance of power between central and local government to recognize that even in the 1930s local authorities were seen as subservient. While, as Young describes, Robson was greatly concerned about this development, Stamp, albeit following Sidney Webb, saw it as not only appropriate but also uncontentious that central government should use the grant system to impose its will on local authorities. It is, to say the least, unexpected to find in a book written to celebrate the achievements of local government a passage such as the following:

> The power of withholding the grant by a Government Department was an economic sanction similar in character to the Sanction of withholding a license for a road vehicle. So long as *ad hoc* offences on the road are dealt with in isolated fashion by magistrates' fines, excesses and anomalies will persist, but as soon as the whole livelihood represented by a license is jeopardised, then we have the most effective engine of correction and restraint. In somewhat the same way, lax behaviour or standards of local authorities came under review. (Stamp in *Municipal Progress*, p. 385)

While no doubt Stamp had specific grants primarily in mind in writing this passage, the notion of using the grant system as an engine

Table 7.4 *Sources of local authority revenue 1930–82, England and Wales*

Year	*Rates*	Total	*Grants* Specific	Non-specific	Total	*Fees, Charges, etc.* Trading Services	Housing Rents
	%	%	%	%	%	%	%
1930	36·5	25·2	20·4	4·8	36·7	25.6	4·9
1935	33·6	27·1	16·9	10·2	37·5	25·4	4·4
1940	33·4	30·2	22·0	8·2	34·6	24.5	4·4
1945	27·1	30·2	23·6	6·6	40·5	28·9	3·6
1950	34·0	34·0	28.5	5·5	29·8	15.5	5·8
1955	32·7	36·1	30·7	5·4	29·0	13·0	8·4
1960	33·3	36·2	10·9	25·3	27·9	10·9	10·3
1965	33·1	36·8	10·4	26·4	26·8	9·2	10·9
1970	26·7	34·6	6·1	28·5	30·0	7·3	11·2
1975	23·5	45·3	10·1	35·2	26·1	2·3	11·1
1980	27·6	52·6	16·0	36·6	19·4	1·5	6·5
1982	31·4	45·1	12·8	32·3	22·1	1·5	8·9

Note: All columns are expressed as a percentage of total local authority revenue. Owing to changes in definitions the figures for 1970 and subsequent years are not exactly comparable with those before 1970

Sources: 1930–75 Foster *et al.* (1980) pp. 147–8; 1980 and 1982 *Local Government Financial Statistics*

of correction and restraint is scarcely the unprecedented assault on the democratic traditions of local government that many critics of the 1980 block grant have claimed (for example, Burgess and Travers, 1980).

Conclusion

The sources of local government revenue over the past fifty years are set out in Table 7.4. That table shows that over the period a progressively smaller proportion of local income has been derived locally (from rates or fees and charges) and a correspondingly larger proportion from central government grant. That change in part reflects the change in the composition of expenditure – from trading services financed by charges to the welfare state services financed by rates or grants. In part it reflects the growth in local government spending relative to GDP, and thus relative to the revenue yielding capacity of rates.

The continuing growth of the grant proportion and the associated weakening of local financial accountability and increased dependence of local authorities on central government have been a source of concern throughout the period, and were a central element of the Layfield Committee's diagnosis of the problems of local finance in 1976. The search for new sources of local revenue has motivated two research studies by the Royal Institute of Public Administration (1956; Hildersley and Nottage, 1968), two Green Papers (Cmnd 4741, DoE, 1971; Cmnd 8449, DoE, 1981), the latter the subject of a House of Commons Select Committee Report (1982), and formed a major part of the work of the Layfield Committee (1976). While it is clear that there are a number of relatively low-yield taxes that could conveniently be given to local authorities, it is now generally appreciated that only a tax of major revenue yield could make any significant impact, and the only such tax that is at all suitable is income tax. For a variety of political and administrative reasons, central governments of both parties have ruled out a local income tax (most recently in Department of the Environment, 1983). The prospects for reform for local government finance, to escape from the present bitterness, therefore appear rather bleak.

Yet the history set out in this chapter suggests that the cause of the problem lies rather deeper than Layfield had thought. It is not the grant proportion but the redistributive nature of most of the services provided by local authorities that invites, or even necessitates, intervention by central government. That intervention is required to maintain reasonable uniformity of standards of services and of local taxes. If some areas, often already poor, have high spending and high taxes and others, already rich, low spending and low taxes, the migration of people and business will ensure that the rich areas get richer and the poor poorer. If local autonomy is to be revived, the extent of redistribution in local government services would need to be reduced. Whatever the other political merits or drawbacks of the free-market liberal view of redistribution through cash transfers combined with charging for services, it would clearly be conducive to greater autonomy in local government. For then central government would take over the troublesome redistributive element (the cash transfers) while local authorities could operate many of their existing services, such as education or housing, as trading services.

Another ten or twenty years of Thatcherite government may well see such a spread of free market liberalism, with local government the unintended beneficiary. There is, on the other hand, no sign of any conscious effort by central government to consider, let alone implement, financial policies to restore the effectiveness and independence of local government. Even such obvious and overdue measures as rating revaluation are delayed for national party political advantage. For as long as central government has no real interest in the success or even the survival of local government as a separate political institution, and for as long as it chooses to regard local authorities as instruments of central policy, the system of local government finance can only stagger from one crisis to another, with technical solutions to particular problems sinking in the conflict of objectives and of purpose.

8

The Rise and Fall of Planning

MALCOLM GRANT and PATSY HEALEY

> The planning and rebuilding of the old residential portions of the city is the task of the next generation. It is perhaps the greatest task our city fathers have ever undertaken. But it is well within their power. If public opinion remains steadily determined to put an end to the slums and prepared to pay the necessary price, if our city councils act with wisdom, foresight, and discretion, we can certainly within a single generation provide for every family a good house in a healthy, convenient, and beautiful city.
>
> Sir Ernest Simon in *A Century of Municipal Progress*, p. 218

Compared to the situation in 1935, contemporary British land use planning represents a substantial and comprehensive set of powers to regulate the land market and to acquire land and undertake development, in order to achieve national and regional spatial, land use and design policy objectives. Overseas commentators writing on the Town and Country Planning Act 1947 which established the main parameters of the present system saw it as revolutionary (Haar, 1951) or disastrous (Dunham, 1964), depending on their attitudes to social policy and private property rights. Yet to the contributors to the original book writing in the 1930s, the establishment of a set of effective planning controls lay far into the realms of idealism. Sir Ernest Simon wrote that 'Planning on a national scale has as yet hardly been thought about' (*Municipal Progress*, p. 214).

Taking stock of the development of a planning system in 1935 could not have been particularly encouraging. The coalition government had not long before succeeded in selling the Town and Country Planning Act 1932 to a distinctly unenthusiastic Parliament, by appealing simultaneously to landlords' self-interest (based on

American experience, planning would build and preserve land values by preventing unwelcome development) and the amenity preservation lobby (planning would prevent urban encroachment, even if this meant destroying present development expectations and hence undermining land values). But the Act fell so far short of providing any solution to the problems of regulatory betterment and worsenment – that is, the redistribution of land values resulting from intervention in the name of planning control – that it forced those councils who could be persuaded to implement the Act at all to over-allocate land grossly in an attempt to avoid any compensation liability. It denied local authorities 'a fair chance of planning effectively in the general interests of the public without incurring undue financial burdens' (*Municipal Progress*, p. 215).

The result was that it in practice served to facilitate the rapid expansion in private housing for sale. This was particularly encouraged around London by the extension of the Metropolitan Line and the building of new roads. Coinciding with continuing depression in agriculture, speculative land-dealing produced an average annual rate of agricultural land-take for urban purposes in England and Wales between 1931 and 1939 of over 25,000 hectares, a figure never since exceeded, even in the boom years of the late 1960s (Best, 1977). This sprawl not only created servicing problems for local authorities but also strengthened the voice of the amenity lobby anxious to conserve rural landscapes and culture and provide access to the countryside for the urban population (Fogarty, 1948).

Land use planning in the 1930s was thus drawn into conflicts over whose interests should be maximized in land allocation but, without effective powers to affect the land market, it could play little role in resolving these. By comparison with today's system, the 1932 legislation was primitive and inherently impracticable. It was parochial in operation, yet highly centralized in control. Despite the egregious optimism of successive annual reports on progress issued by the old Ministry of Health taking as their yardstick the figures of resolutions passed and schemes submitted and approved, the truth was that it lacked the basis of any firm national policy and it impinged remarkably little on landowners and developers. Planners were left to pursue the same philosophy espoused by John Burns, as President of the Local Government Board, as long before as 1909, that planning would succeed best as a negotiated, co-operative venture between landowners and local authorities, rather than as a firm regulatory machine. If authorities attempted to go beyond this, for example to limit the polluting effects of local industry, central government intervened to restrain them (Gunby, 1984).

The sceptic might remark that little has changed. Despite the postwar emergence of a highly sophisticated planning system, which managed at last to sever the regulation/compensation link, which provided for the adoption of forward-looking policies and which conferred the powers and even, at times, the money to pursue them, there are many features which still look the same. The most notable is that the national economy is again in a state of acute depression, and planning for growth has had to give way to the management of urban decline. Yet there is still the same battle that was fought in the 1930s – albeit from different baselines – over the release of land for private housebuilding. Negotiation rather than unilateral control still characterizes the processing of most larger-scale development proposals, and planning is seen to follow the market – even to be manipulated by it in the name of 'planning gain' – rather than to overturn short-term market dictates in favour of longer-term objectives. Perhaps even more fundamentally, the early idealism of planning as based on a general consensus as to what was decent and good has given way to a more general ambivalence.

Such superficial comparison is obviously misleading. The postwar land use planning system has radically altered the operation of the land market, to the extent that investment decisions now assume the existence of policies to contain urban development and to limit land allocations for particular activities within carefully analysed projections of demands and needs. There is little support for the abolition of land use planning. Even the Adam Smith Institute (1983) did not suggest this in its recent review of planning and local government. What has shifted in the postwar period is the balance of powers and resources available within the system and the interests the system has been encouraged to emphasize. One symptom of this process is the frequent legislative tinkering and judicial review which planning law has undergone. Nevertheless, the basic outline of the system has remained surprisingly stable. State regulation of private land use and development rights exercised through administrative discretion, guided by publicly prepared plans, supplemented by a range of opportunities for objection and review, and supported by powers of land acquisition and development, still provide the main tools of the system. Local authorities still play the main role in exercising these powers though carefully scrutinized by central government.

The true reason for the constant fluctuation of policy and amendment of the machinery of the planning system goes deeper than the legal structures. It is an inherent consequence of the breakdown of Lewis Silkin's short-lived but effective postwar consensus as to the objectives of planning. It was too delicate and tentative to be spelled out in the legislation, and even to this day judges and politicians struggle with the deliberately loose language inherited from the 1947 Act. The flexibility of development control, for example, is enshrined in a provision which allows the development plan to be displaced in decision-making by any other material consideration. Silkin's Act attempted, therefore, to make up for a want in substance by providing procedural protection to propertied and bureaucratic interests. There was to be close overview of local authority activity through an ambiguous mix of quasi-judicial, political and administrative measures, and this left the system particularly vulnerable both to longer-term changes in central–local relations, local government structures and judicial review, and to shorter-term political instability in which elaborate procedures served to conceal and legitimize highly subjective or crudely ideological choices.

The principal function to which the planning system has been put has therefore become that of mediation between competing interests in a loosely structured political system sharply hedged about by legal procedures. It follows that any qualitative assessment of planning's achievements depends upon the extent to which the values to which the observer subscribes have been advanced. And there are no easy quantitative indices of performance.

To understand how the planning system has changed in the past fifty years it is therefore necessary to examine how the state has responded to the changing economic and social pressures acting upon it as reflected in land use and development issues, and how these have affected the political and institutional relationships through which the system operates.

The State and Land Use Change and Development

It is widely recognized that as urbanization proceeds, the state is increasingly drawn in to manage land use change and development.

From an initial concern with health hazards and mass housing, by the 1930s planning became involved in the problems of servicing development and mediating between conflicting interests over how land should be used. In other words, the state was required to intervene to facilitate individual activities (notably housebuilding) and organize spatial externality effects. The objectives were not just to avoid pollution but to make urban development more efficient and to provide for the various demands and needs for land and buildings. However, these problems were only clearly appreciated by a few at that time, and the main issues which preoccupied the small band of officials implementing the Planning Act 1932 were the reservation of land in new development for open space and public purposes, controlling densities and prevention of ribbon development.

The experience of national planning during the war, the large-scale destruction of the physical environment as a result of bombing, and the ideology of a Labour government committed to a national economic and social welfare strategy produced a convergence of ideas about the proper spatial arrangement of towns, and spatial planning within a national strategy, which led to the 1947 Act. Of particular importance was the high priority given to the protection of agricultural land, consequent upon the postwar commitment to ensure a high level of home food production. The location of development was to be regulated within the context of development plans which were to realize the spatial and physical implications of national policies and facilitate major state programmes of urban reconstruction and new town building. Meanwhile, speculation in development land was to be overcome by a special development charge aimed to hold the market to transactions at existing use values. The main priorities for land use planning were to be first, to contain urban development to reduce servicing costs and protect agricultural land, reinforcing the early restrictions under the Restriction of Ribbon Development Act 1935 by allocating land for development in statutory plans in accordance with population and growth projections. The second priority was to redevelop war-damaged areas and congested town centres. Thirdly, slums were to be cleared to relieve inner city congestion. Finally, reasonable quality in the built environment was to be ensured, with particular emphasis on the control of advertising.

There was considerable discussion in the framing of the 1947 Act as to which level of government was to carry the main responsibility for it. Central government officials were sceptical of the competence of the 1,441 local authorities responsible for planning in the 1930s. There was also a suspicion that many local councils were far too deeply associated with local landowning interests to be able to exercise an effective role in containing urban development (Cullingworth, 1975). The strongly centralized Labour government was already transferring tasks from many local authorities to national public agencies in energy, health, social security and communications. The idea of planning commissions divorced from local government was floated, and received support from central government officials. However, the local government lobby, building on their experience with the 1932 Act, were successful in ensuring that the main role in implementing the new system stayed with the 145 counties and county boroughs.

The 1947 Act as originally envisaged was given little time to produce effects. The Conservative administrations of the 1950s swiftly repealed the development charge and had little interest in addressing questions of spatial policy apart from the regional issues of the distribution of economic activity. Local authorities who, with the help of consultants, produced their first round of development plans in the early 1950s were thus left to implement these with little interference from the centre. Local and professional preoccupations at this time were town centre redevelopment and slum clearance in the cities, and managing the balance between accommodating development needs and minimizing land-take on the urban periphery. Meanwhile, new town corporations proceeded with building a considerable number of new settlements. Conservative legislation gave special powers for planned town expansion agreements between local authorities, as a more acceptable approach to relieving urban congestion than the new town mechanism with its overtones of land nationalization.

But the development pressures of an expanding economy and population soon forced central government to recover an interest in national spatial policy. Having promoted the Green Belt as a containment device in 1955, central government was faced by the end of the decade with an all too enthusiastic response from local authorities. As a result the amount of new housing land allocated in development plans appeared to be out of step with the demands of the housebuilding industry, which vigorously expanded production encouraged by government policies of increasing home-ownership. Throughout the

1960s, governments of different political complexions pressurized local authorities into allocating more housing land on urban peripheries. But the political difficulties of doing this helped to support the policies of high-rise high-density inner city redevelopment with the disastrous results so evident by 1985. When the Labour government returned to the question of recouping community-created development gain, it was in conjunction with a national Land Commission, the specific brief for which was to ensure adequate land was available for development. It did not hesitate to force the hand of local authorities, and had few friends when it was abolished in 1970.

During the 1960s, then, central government was prepared to moderate the local articulation of planning policy in order to facilitate the housebuilding interest in particular. Yet at the same time, it steered a change in the development plan system through Parliament which devolved the approval of detailed policies and proposals to the local level. In part this was an acknowledgement that the centre did not have the knowledge or the political sensitivity to assess the reasonableness of many policies. It was also hoped that the distinction between strategic and local policies would assist the centre in focusing more clearly on the matters it *was* interested in.

But as this two-tier planning system was put into operation in the 1970s central government intervened in local planning policy more rather than less. There are many reasons for this. Local government reorganization in 1974 split the planning function between counties and districts, and central government was drawn in to arbitrate in inter-authority disputes. The separation of strategic and detailed issues proved difficult to sustain. In addition, the deepening recession has led central government not only into facilitating housing development and the property sector, but also into encouraging industrial and employment-generating development of any kind. It has thus sought to change established strategic policies and to secure their translation into locally implemented decisions. Finally, and particularly in the case of the post-1979 Conservative administration, party ideology has blamed local government bureaucracy for inefficiency and outdated policies. The then Secretary of State for the Environment, Michael Heseltine, accused planning authorities of locking jobs up in filing cabinets, and started to monitor their performance in processing applications.

The centre has thus put its weight behind facilitating particular interests within the complex balancing act between multiple interests in land which is at the heart of the planning system. It often appears as if it leaves local authorities to do the difficult job of striking the balance, while criticizing them for 'delay' or over-restrictiveness. Yet from time to time the centre itself is caught in the cross-fire of competing interests. The recent attempts by central government to produce circulars on housing land and Green Belts has been a particularly uncomfortable episode for a rather naïve Secretary of State.

In the thirty-eight years since the Town and Country Planning Act 1947, central government has been unable to determine unambiguously what its proper role in the planning system should be. Meanwhile local authorities have greatly increased their competence through the expansion in size and training of their staff. Yet behind the shifting sands of central–local relations in the planning system have been important changes in the demands, needs and organization of the various interests the planning system attempts to balance.

Facilitating and Balancing Interests

With the exception of a few major industrial firms and certain landowners, most land and development issues in the 1930s arose as a result of the activities of firms and pressure groups closely linked to local authority areas. The local articulation of planning policy thus involved maximizing and balancing the interests of its own constituents. In the postwar era, this situation has changed. The various sectors of the economy have increasingly been organized into large-scale national and multinational entities often with few ties with the localities which accommodate their activities. In constructing planning policies local authorities are thus faced with devising a compromise between large-scale productive operations and local community demands and needs.

Central government has consistently protected certain production interests. The privileged position accorded to the agricultural sector in postwar planning was in part a necessary continuation of a wartime commitment to exclusive home food production, and in part a reflection of the philosophy of the Scott Report (1942) that agricultural land should be safeguarded from urban encroachment. Operations 'requisite' for agricultural purposes were therefore made exempt from planning control, and over the succeeding years the strength of the agricultural lobby ensured that the exemption became the more firmly entrenched the more agriculture itself changed from the type of undertaking to which the exemptions related: that is, a predominantly small-scale pastoral operation undertaken on small units within large estates. It meant that as a general rule there was to be no planning control over the introduction of intensive livestock farming, no control over changes in agricultural uses such as the giving over of productive land to recreational 'horsiculture' and other more remunerative uses, and no control over operations such as grubbing up hedgerows, draining ponds, filling gullies and ploughing up moorlands, which were to transform the appearance of the countryside.

Much the same degree of protection has been accorded to mining operations, including the expanding open-cast mining industry. As the minerals industry has become more highly centralized, it too has been accorded special status in planning procedures, with development control organized at county, not district, level and the aggregates working parties to negotiate with counties the amount and location of land to be released for mineral working.

By contrast, industry was not accorded any special position in planning procedures. Until the 1970s the general view was that, with a few exceptions where firms were sufficiently powerful to influence central government, industry could be deflected to depressed regions or better located zones within urban areas without cost. The main policy thrust until recently was thus towards regional redistribution. The 75 per cent de-rating of industry which the government had conceded in 1928 as a stimulant to industrial activity was withdrawn progressively in 1958 and 1961, and a new system of industrial location controls and incentives introduced in line with the Barlow Committee's recommendations (1940). The impact of the incentives, which included regional development grant and regionally biased sectoral aid programmes, was more positively felt than the controls built into the planning system. Refusal of an industrial development certificate meant that planning permission could not be granted for development, but only a small proportion of industrialists responded by choosing to relocate in the regions. For the most part the controls had a dampening effect on new industrial development, even though in the buoyant areas – most particularly the West Midlands – its impact was not fully felt until the collapse of traditional industries showed up how imbalanced the industrial base had become.

Outside the depressed regions, industrial development tended not to be a high priority for either central or local government, and more particularly so after the emergence of an articulate and well-organized conservation and amenity lobby in the mid-1960s. It was rare, until it was forced upon them by the unemployment crisis of the late-1970s onwards, for local authorities generally to adopt a corporate approach to attracting and promoting industry. When it did come, it was a process very largely of local initiatives despite the involvement by central government in the partnership and programme areas declared under the Inner Urban Areas Act 1978 and the enterprise zones, free ports, derelict land clearance and urban development areas. Local authorities for their part operated through policies of selective (and therefore often highly discriminatory) relaxation of planning controls, by securing industrial land supply and by underwriting redevelopment, assisting small businesses and promoting employment generally. But this has come at a time when the progressive withdrawal of central government rate support grant, coupled with expenditure cuts and penalties for spending beyond centrally imposed targets, has made local authority marginal expenditure bear substantially more heavily on ratepayers than before. The effect has been not only to limit spending on economic development but also to impose new corporate pressures to make up the shortfall by distorting regulatory controls, such as planning, which are not expenditure-linked but which realize development value in land, to underwrite development.

Commercial property development and the housebuilding industry have by contrast been greatly facilitated by the planning system, although until recently no special mechanisms for negotiation between property developers and planning authorities were encouraged. As many studies have shown, in the 1950s and 1960s such developers were typically locally based and had ample opportunity to influence their planning authorities (Marriott, 1969; Ambrose and Colenutt, 1975; Saunders, 1979).

The growth of the commercial sector was rapid after the war, fuelled by the expansion of credit of the 1950s and 1960s. Local authorities helped with compulsory purchase in land assembly and land supply for town centre redevelopment and often accepted highly disadvantageous financial terms, offering long leases at rents which were historically low and non-reviewable, in order to attract the development expertise and finance they required. They tended also to be vulnerable to manipulation by developers seeking planning trade-offs, such as relaxed plot ratio standards or off-street car parking requirements. George Brown's office ban in 1965 guaranteed demand for office blocks already constructed or in the pipeline and pushed up rents in the metropolitan area and home counties. The effect of a combination of high rents and low historic yields was to boost capital values enormously. The spectacular profits which were forthcoming from these ventures encouraged the financial institutions to divert even greater resources to financing property development. The collapse of the property boom in 1973 led to a serious crisis in the major undiversified property companies and the secondary banks (see Barras, 1985). One consequence was the increasing centralization of the property sector, closely linked with financial institutions.

Retail development was one branch of the commercial sector which was almost exclusively own-use, though increasingly centralized. It caught local planning authorities on the hop in the mid-1970s by shifting to out-of-town shopping. It was a topic to which the formal processes of statutory planning were ill-suited, and most of the early hypermarket proposals met strong local commercial resistance, only to be allowed on appeal by the Secretary of State as departures from the urban containment policies of the statutory plans. Sites chosen for hypermarket development tended to be either agricultural land on the urban fringe, or low value inner city sites, and the very substantial jump in land values which occurred when retail permission was granted provided a windfall benefit from which many of the big retailers were prepared to offer sizeable planning gain inducements to local authorities. The shift in shopping patterns thus became something which local authority planners were either unable or unwilling to counter.

The housebuilding industry in its turn experienced a steady shift towards larger-scale organization, with the rise and increasing dominance of the volume builder, often linked to industrial, commercial and financial conglomerates. Because of the significance of development gain in the profit structure of housebuilders (Ball, 1983), the planning system has been consistently under pressure to realize that most profitable of circumstances, a general restriction on housing land supply, coupled with the allocation of a continuous flow of sites so that land bought at agricultural value will eventually realize development value. As we have shown, housing land availability has been a preoccupation of central government since 1960. The industry's dependence upon a guaranteed land supply was reinforced by rising interest rates and development land tax (DLT), which was introduced in 1976. DLT proved a largely voluntary tax and brought in little revenue, but the avoidance techniques widely adopted by landowners relied heavily on phasing land release. The tax therefore made it more difficult for housebuilders to maintain the large land banks of the 1960s and early 1970s and central government took an increasingly strong line with local authorities to bring land into development. In the 1960s local authorities were merely encouraged to allocate land. By the early 1970s the emphasis was on ensuring that it could be serviced. By the 1980s marketability had become a major criterion (Hooper, 1985), and authorities were encouraged to develop joint working arrangements with the House Builders Federation. But by this time the large-scale volume builders had broken away from the federation, in order to develop their own channels of influence to central government. In this industry too the trend to large-scale organization has created pressures for more corporate working relations between the industry and the planning system, focused increasingly at the national rather than the local level.

As the centralizing tendencies within economic organization have produced an industrial and commercial structure increasingly remote from close ties with specific localities, the public, as citizens and consumers of the environments regulated by the planning system, have become sceptical of the capacity of planning authorities to take adequate account of their interests. In 1947 it was assumed that local authorities moderated by central government could construct in an objective way the appropriate balance between interests which would serve the public interest. As production interests have challenged the way local authorities have done this and have put pressure on central government to alter the way planning authorities have constructed the public interest calculus, so groups representing local residents have turned to political action in the form of lobbying pressure groups to protect their interests. The most powerful of such groupings have been those who have linked a specific concern to protect a local environment from further development with the wider environmental movement which achieved political momentum from the late 1960s.

The impact of the growing amenity and conservation movement within the planning system started to be felt by the mid-1960s. The success of the Civic Trust in promoting urban conservation, and in gaining firm parliamentary support for it in the Civic Amenities Act 1967, was to have an enormous effect on planning ideology. Planning authorities on the expanding urban periphery, who in the 1960s had been energetically encouraging the remodelling of their built environment to make it more appropriate for modern transport and commercial operations, had by the mid-1970s become staunch defenders and extenders of Green Belt and other restraint policies. The Council for the Protection of Rural England (CPRE) and the Civic Trust provided national lobbying machinery to support local campaigns. Inner city residents also voiced objections to the destruction of their environments consequent upon city centre redevelopment and slum clearance programmes. The benign paternalism of postwar municipal socialism in many cities was increasingly challenged by activists seeking more sensitive solutions to local environmental problems and by the realization that those in inner cities were disadvantaged largely through national changes in social and economic organization. Their problems therefore required action at the national level. Rather than confront this problem politically, governments from the late 1960s allowed themselves to be preoccupied with the so-called inner city problem, as if local authority incompetence and inner city residents were responsible for it.

In general, where central government's role in the planning system has increased, it has been in response to the increasing scale of organization of the production sector and the parallel national lobbying of amenity and conservation pressure groups. Reflecting the established sectoral organization of central government, it has exercised this role by pressing local authorities to give more attention to particular interests. But it has given them little advice on how to order priorities in specific localities in a way which is both economically efficient and politically sustainable. As a result, central government interventions have increased the inconsistency in the operation of the planning system which many of the major interests in land and development complain so much about.

The Procedures and Politics of the Planning System

A central and still largely unresolved problem within the planning system has thus been how to design a set of institutional relationships which allows the construction of policies and proposals to facilitate individual sectors, organize spatial externality effects and resolve interest conflicts in a way which is accepted as legitimate by all parties concerned and which provides reasonably stable conditions in which long-term investment decisions can be made.

In the 1940s it was assumed that the problem could be resolved by the application of technical expertise, supplemented by procedural protection for key interests such as agriculture, and the procedures for objection and inquiry to ensure that individual property-owners could get their interests properly addressed. But the availability of trained planners remained limited until the 1970s. Nor did their training necessarily help them to address the task. The traditions of town design propagated by Keeble (1959) in the 1950s paid little attention to the economic, social and political processes which produced the physical environment. The regional geographers moved into the planning field in the 1960s with greater sensitivity to the causes of spatial change, but little understanding of the planning system or of political processes. The planning profession itself absorbed the principles of corporate planning and rational decision-making in the 1960s, but with little appreciation of how to relate these to the land and development issues that still had to be addressed, and with an arrogant view of the relationship between technocrats and politicians. As local government became increasingly politicized from the late 1960s onwards, planners were attacked both for over-ambition in taking too much responsibility for articulating policies, and for failure to deliver an adequate built environment (Healey, 1983).

Faced with increasingly vociferous lobbying from powerful production interests and amenity pressure groups, many planners in the early 1970s turned to the notion of public consultation as a way of constructing planning policies. Public participation was afforded official recognition by the Skeffington Committee Report (Ministry of Housing and Local Government, 1969) and mechanisms for formal participation in plan participation were included in the Town and Country Planning Act 1968. Public participation offered an alternative view of the public interest and by exploding what was left of Silkin's grand consensus, laid bare the conflictual nature of planning decision-making and its mediating role.

The increasing politicization of local government served to reinforce this perspective. Policy was susceptible to changes not only in overall political control of an authority, but also within party groups. The growth of ideological politics charted elsewhere in this volume had a direct impact upon planning. It resulted in the harnessing of planning controls to crude ideological ends, such as in attempts on the left to impose restrictions on speculative office development, private education and private health care facilities, and (on the right) to prevent growth so as to protect existing property values or commercial viability, to frustrate central policy as under the Community Land Act and to facilitate the disposal of assets. But it also had the broader effect of creating greater policy differences within and between local authorities, and it undermined the continuity of policy and consistency between levels of government which statutory plans implied.

Public participation and political ideology highlighted the conflicts of interest within the planning system, and the importance of the mediating machinery within the planning system became evident. In recognition of this, the Local Government, Planning and Land Act 1980 and subsequent advisory circulars encouraged local authorities to reduce the time taken on technical work and public consultation, and to use the public inquiry as the main mediation device. It is also used to resolve disputes between the two tiers of local authority, and as a vehicle (through appeal and local plan inquiries and the more streamlined variant of the structure plan Examination in Public) for central government to amend local authority policy and decisions.

Planning through inquiry may appear reasonable to the interests concerned. But its forensic style makes it a cumbersome procedure to use where complex technical issues and major policy questions are involved, as the experience of the public inquiries into major public investment projects has shown. By focusing on specific objections it also tends to atomize issues, rather than to interrelate them. Nor does it necessarily ensure that the policies and proposals finally adopted in plans and the decisions in planning appeals are actually implementable (Healey *et al.*, 1985).

For planning in contemporary conditions is not just about providing a land allocation framework. Land has to be cleared and serviced if development is to proceed. In contemporary conditions, with policies of public expenditure restraint, and with most production interests having considerable difficulty in putting together financially viable development projects, both public and private sectors are pushed into a position of mutual dependency. This dependency is more than the negotiation and bargaining characteristic of the relations between planning authorities and green field site developers in the 1930s, 1950s and 1960s. In such situations developers used bargaining to influence land allocation policy and attract public sector services to their sites. The current relationship is more akin to the nature of the town centre redevelopment partnerships of the 1960s, where public sector powers of land acquisition and highway relocation were married with developers' concern to reassemble plots into appropriate parcels for modern commercial activities.

In contemporary conditions, the public sector seeks the benefits of job-creating investment, environmental improvement and payment for infrastructure, while developers want land assembly, infrastructure provision, land clearance, a flexible attitude to land allocations combined with protection for property values. Central government has encouraged such co-operative relationships through exhortations to joint working arrangements. It has also provided a variety of financial measures through urban programme moneys, urban development grants and the revised derelict land grants, which offer subsidies within partnership arrangements. It nevertheless shows little understanding of the already widespread phenomenon of planning gain, in which local authorities consciously trade planning objectives for financial gain. It may be the provision of a civic amenity, such as public open space, recreational facilities or footpaths, or it may take the form of dedication of land, the gifting of money or the discontinuance of an authorized but non-complying use elsewhere which the local authority lack the resources to buy out under normal discontinuance procedures. From the local authority's viewpoint, the current target and penalty system of expenditure controls means that the arrangement may have a financial benefit far exceeding its actual price because it relieves them of an item of marginal spending at a high rate of grant-loss. For the developer, the price is almost invariably either passed back in full to the landowner as a diminution of the development value otherwise accruing to the site, or offset against development land tax liability.

The Future

There are presently two clear and related trends in planning. The first is the trend towards reducing the scope and influence of planning at the local level. It is underpinned not only by the dual forces of managerial efficiency and privatization, but also by the growing centralization of government. One of the reasons advanced for the proposed abolition of the GLC and the metropolitan counties is that their strategic planning role is largely irrelevant; while financial controls over the remaining authorities have limited their ability to devote resources not only to control but, more importantly, to the implementation of plans. The second trend is that of partnership between public and private sectors in land development. Partnership has obvious attractions as a general concept with its symbolism of consensus and sharing. But in practice it raises fundamental difficulties which have not been clearly articulated or even always recognized. Local authorities make particularly vulnerable partners. The opportunity to negotiate from strength which the Community Land Act offered even in its initial phases has been lost. The local authority role is confined principally to underwriting development which the market would not otherwise produce, but without the necessary financial resources to do this on a large scale. The asset which they bring to the partnership may either be land which they now lack the resources to develop directly, or the promise of selective and discriminatory regulatory action, such as the use of compulsory powers against existing landowners to assemble land for the development or the relaxation of a planning control which will enhance the value of the finished development. Deals are negotiated in secret with development plans used, not for guiding and controlling growth, but as a basis for pragmatic negotiation to facilitate particular developments and to favour particular interests. Public participation is squeezed out, and policy conflict internalized within the local authority.

These two trends have been actively encouraged in the early 1980s. However, it is debatable whether they will persist in the long term. Neither trend facilitates an adequate response to the serious problems of recycling land in older urban areas and striking a sustainable balance between the needs of the economy and strongly held social and environmental values. Planning may have been under attack in recent years. Yet the call of the author of *Municipal Progress* quoted at the start of this chapter could well be reiterated today. There are important tasks to be done in reconstructing our cities. City councils can act with wisdom, foresight and discretion. The challenge for the next fifty years is the same as for the last: to reassert the ideology of longer-term planning over the short-term ideology of propertied interest and development profits. It is a process which requires not just legislative tinkering but a more fundamental reconsideration of the principles governing intersectoral and central–local relationships and the objectives to be pursued through them.

9

The Nationalization of Housing Policy

ALAN MURIE

> One vital point that must be remembered is that we cannot get the families of the lower-paid workers out of the slums unless we build new houses for them to let at rents within their means. If we are to house the present slum population in new houses of the Tudor Walters type, we must face the fact that very substantial subsidies will be required.
>
> Sir Ernest Simon in *A Century of Municipal Progress*, p. 218

The View from 1935

By 1935 housing policy had moved beyond the health of towns question and sanitary reform. The legislation of the 1920s had added exchequer subsidies to the powers of local authorities to build houses. The intention to build 'homes fit for heroes', the threat of civil unrest, shortages of labour and high construction costs had led to major innovations in policy and a major public investment in housing. The Housing (Financial Provisions) Act 1924 in particular had 'estabished the local authorities as part of the permanent machinery for providing working-class houses' (Bowley, 1945, p. 40). By 1935 some 950,000 local authority dwellings had been built in Great Britain. Substantial reconditioning of dwellings had also had an impact on housing conditions, and Bowley (1945, p. 147) suggested some 300,000 houses had been made fit for human habitation through such activity in each year between 1919 and 1930. Slum clearance was beginning to expand following the Housing Act 1930.

Nevertheless Sir Ernest Simon's verdict on a century of progress in the area of housing and civic planning was mixed. The greatest achievement of our city fathers had been 'the conquest of insanitary conditions'. The standards of new dwellings had 'now reached its goal' and the contrast between the hovel of a hundred years ago and the Tudor Walters cottage on a new housing estate 'must be seen to be believed'. In Simon's view it was difficult to see how this standard could be improved. However Simon identified one very definite feature – the failure to deal with the housing of the poor. The low-paid worker with a family still lived in overcrowded slums and the price of new buildings to rent or buy was more than such families could pay. The old slums had only begun to be cleared. Some had been improved by reconditioning (and all by the provision of sanitation) but millions of houses in use remained below the acceptable standard. State intervention to improve the standard of housing had forced up costs and therefore rents. Those with low incomes could only benefit from increased standards with the help of subsidy.

Sir Ernest Simon's view of the housing situation was coloured by the circumstances of 1935. In particular, falling interest rates and low building costs had created the climate for an unprecedented speculative building boom. The year 1935 marked the high point for private sector building in the interwar years and more than offset the decline in public sector building. Total new construction was at its *highest recorded level and was still rising*. These favourable and optimism-creating circumstances have been summarized by Headey as follows:

> In terms of impact on housing conditions and costs the private enterprise boom of the 1930s dwarfed deliberate acts of public policy. An average of 263,000 houses a year were built from 1935/6 to 1938/9. This was over 100,000 up on the previous four year period and 130,000 more than the annual rate for the previous decade.
>
> Private enterprise boomed and consequently diminished the overall severity of the 1930s depression, because all the incentives were right at the same time. Generally rising incomes and the fact that the age distribution of the population was such that a lot of young couples were marrying and setting up homes produced conditions of rising effective demand. Mortgage funds were abundant because the building societies represented a better invest-

> ment than the stock market. Last, in the early part of the decade, building wages and the cost of materials fell, and although they rose later, house building remained a highly profitable enterprise. (Headey, 1978, p. 135)

The impact of these record building rates was identifiable in the rapid fall in levels of overcrowding and occupancy rates and the elimination of crude housing shortage. In the late 1930s 'the total of dwellings was probably slightly greater than the numbers of households for the first time in living memory' (Department of the Environment, 1977a, p. 22).

The achievements of the fifteen years of housing policy which Simon referred to were considerable but were the product of favourable circumstances for speculative activity as well as of public investment. Indeed the achievements of public investment fell far short of early targets. Subsidy levels were not sufficient to bring rents down to levels which lower-income families could afford. Even the period 1919–35 was marked by reductions in public subsidy, in local authority building and in dwelling standards. The subsequent history of housing policy has tended to reproduce the same features in state activity (cuts in subsidy, investment and standards in periods of economic and public expenditure crisis; a concern to encourage the private sector and hold back municipal provision) but not the favourable circumstances for speculative activity. In 1985 while public sector investment and subsidy is at an all-time low private investment has not taken-off. Government has been able to cut back public activity but not to create the climate for increased private sector activity. Concern about housing conditions and shortages is increasing in this environment and the connection between housing problems and poverty remains apparent.

Housing Policy Changes 1935–85

The favourable circumstances influencing Simon's view were brought to an abrupt halt by war. As Cullingworth (1966) has stated:

> In 1939 there was a rough balance between the numbers of houses and families. The virtual cessation of housebuilding during the war, bomb damage and destruction, and a population growth of a

million, resulted in a housing shortage of enormous dimensions. (p. 29)

Housebuilding between the wars had improved standards but it had not ensured comparable basic standards for all households. Available statistics on amenity provision and sharing demonstrate how much remained to be done. The war between 1939 and 1945 made this situation dramatically worse through bomb damage, evacuation and homelessness and the redeployment of labour and materials away from housebuilding and repair. Some 3,745,000 different houses in the United Kingdom were either damaged or destroyed. A million people in England and Wales were living in dwellings which had been, or but for the war would have been, condemned as slums. Eight million people were living in damaged houses which had received only first-aid repairs and many were living in overcrowded circumstances (Titmuss, 1950, p. 411). A series of enactments in 1944 and 1945 tried to deal with this situation by empowering local authorities to erect temporary homes, to purchase land and building materials and by extending the housing subsidies payable under the Housing (Financial Provisions) Act 1938 to support building for general needs.

Following the general election of 1945 the main direction of policy was established. In comparison with the position in 1919 policy was more strongly dependent on local authority activity and wartime controls on the private sector were maintained. The Labour government's Housing (Financial and Miscellaneous Provisions) Act 1946 increased the exchequer subsidy to £16 10s per house for sixty years. A rate contribution of £5 10s was required in addition. This trebled the money value of subsidies compared with 1939 and provided an effective stimulus for local authorities to build. The government was encouraging high standards of building and with general price inflation these subsidy levels were necessary to hold rents at about 10s per week (Merrett, 1979, p. 242) in spite of low interest rates and the availability of cheap loans to local authorities from the Public Works Loans Board (PWLB). Other measures, and particularly planning legislation, speeded up land acquisition and clarified compensation terms in the event of compulsory purchase and set these at existing use value rather than market value, thus taking a potential cost off the Housing Revenue Account.

The 1946 Act and related measures sparked off an unprecedented rate of building of high-quality housing for the working class. Speculative building 'for the purposes of supplementing the main housing programme and not for diverting building labour and resources' was intended to make up at most one-fifth of total housebuilding in each local authority and where authorities were exceeding this level, central government imposed restraints. In addition a general instruction remained that licensed building must go, and be seen to go, to people in need of homes. The upper cost limit which determined whether a licence was required set a level which covered all new private dwellings.

This priority for houses to let implied a very different role for local authorities than had existed before 1939. Aneurin Bevan as Minister of Health in 1945 stated:

> Before the war the housing problems of the middle-classes were roughly solved. The higher income groups had their houses – the lower income groups had not. Speculative builders, supported enthusiastically, and even voraciously, by money-lending organisations, solved the problem of the higher income groups in the matter of housing. We propose to start at the other end. We propose to start to solve, first, the housing difficulties of the lower income groups. In other words, we propose to lay the main emphasis of our programme upon building houses to let. That means that we shall ask Local Authorities to be the main instruments for the housing programme. (House of Commons, 1945)

By 1948, as local authority building completions increased and private completions remained low and even declined, local authorities were playing the dominant role intended for them. Local authority completions (excluding prefabricated and other temporary dwellings) had grown from 3,364 in 1945 to 190,368 in 1948 compared with 576 in 1920, 86,579 in 1922, 120,492 in 1928 and the previous high point of 121,653 in 1939. However this very rapid recovery was not sustained and economic crisis led to cuts in activity. The vulnerability of housing performance to economic crises and the absence of clear target figures for housebuilding represented a failure to plan in the way intended.

In 1949 a new Housing Act introduced for the first time grants for improvement or conversion of dwellings in the public and private sector. The Act also dropped reference to local authority provision as provision of working-class housing. Consistent with Bevan's objection in 1945 to segregated, twilight villages, municipal housing ceased to be restricted to 'the working class'.

The thirteen years of Conservative government from 1951 saw housing activity shifting as a result of more than economic crises. The general direction of policy involved the recreation of market conditions and restriction on local authority building. The relaxation and removal of the building licence system, and the encouragement of private building was initially accompanied by higher subsidies (under the Housing Act 1952) and higher targets for council house building. New public sector building peaked in 1953 and 1954 but after 1954 the role of the local authority was redefined (to clear and rehouse the slums) leaving the general housing need to be met by private enterprise. Public sector building declined between 1954 and 1959 and the growth of private output was not sufficient to offset this. While a Labour government in 1964 embarked on a new housing drive in which local authorities were to be equal contributors the ensuing building boom was short-lived and public sector completions fell away after 1968. The account of housing policy after 1954 could focus on building rates, changing subsidy systems and variations in emphasis on general needs or slum clearance building. Indeed the revision of subsidies in 1956 (reduction in subsidy for general needs building, rate fund contributions optional, subsidy structure encouraging high-rise building); in 1967 (more generous subsidies to local authorities); in 1972 (introducing a deficit subsidy and removing discretion over rent charges); in 1975 (providing new arrangements to replace those under the 1972 Act); and in 1980 (reverting to a deficit subsidy) are important elements and especially latterly have been major sources of conflict between central and local government. What has not been achieved is the level of building Simon envisaged. Simon (1935, p. 218) considered that:

> If we continue the present rate of building we shall rehouse one-fifth of the population every decade. In thirty years we shall have rehoused three-fifths of the population, or, including what has already been done since the war over four-fifths of the population will be living in postwar houses.

But in 1975 only 44 per cent of the stock was of postwar dwellings and 33 per cent was of pre-1914 dwellings. Simon's optimism after fifteen years of municipal housing has a more hollow ring after a further fifty years. The task has not proved as simple as he suggested. Levels of investment in 1985 are a long way below those needed to achieve the modernization of stock envisaged by Simon. Government has ceased to consider housing provision in the light of estimates of need and 'what the country can afford' has become purely a public expenditure rather than a housing calculation. Subsidies and new building have also come to be dominated by considerations of tenure. In the interwar years the growth of owner-occupation progressed rapidly through new building and sales of privately rented properties. But as the private rented sector has declined so its contribution to the future growth of home-ownership has declined. Maintaining the rate of growth towards a property-owning democracy has increasingly involved halting the growth of council housing and selling council houses and encouraging local authorities to sell land and adopt various low-cost home-ownership initiatives often involving low standards as well as subsidizing home-ownership.

But what has happened in housing and the role of local authorities in housing goes much further than an account of subsidy and building would imply. Six important dimensions can be referred to. Firstly, housing programmes have been a consistent victim of economic crises. Housing has always been a victim of economic fluctuations. But the inflation and high interest rates of the 1970s and the economic orthodoxies of the 1980s have eaten into the housing programme more dramatically than in the past or than improvements in the housing situation would imply. By 1981 new local authority housebuilding had fallen to its lowest peacetime figure since 1926. By 1983–4 exchequer subsidy had fallen to £290 million (compared with £1,423 million in 1980–1, £313 million in 1970, £120 million in 1960 and £72 million in 1950).

Secondly, housing policy has increasingly come to be dominated by tenure issues. The preoccupation with building targets has been replaced by a preoccupation with the growth of home-ownership. In spite of periodic attempts to revive it private renting has declined consistently since 1935. Indeed its decline has been speeded by the fiscal and other encouragement of owner-occupation. The encouragement of home-ownership has been pursued by both Labour and Conservative governments which have converged in competing for the owner-occupier vote. As home-ownership has been seen to be electorally important so it has been seen to be the natural tenure.

Policies which would interfere with the expectations of home-owners have come to be regarded as a liability. In contrast the expectations, rents and rights of households in other tenures have not been sacrosanct and ideas of tenure neutral policies have made little progress. Local autonomy has been overridden when it has appeared likely to delay progress towards home-ownership and a property-owning democracy.

Thirdly, hidden 'demand-side' incentives to owner-occupiers have not been regarded as public expenditure but have increasingly affected the housing market. The biggest growth area in housing policy since 1951 has been financial and other encouragement to owner-occupiers. Some of this has come through direct legislative action in the option mortgage scheme introduced in 1978, and the extension of improvement and repair grants following their introduction in 1949. But more important has been the increasing value of fiscal relief associated with house purchase. Such tax reliefs existed between the wars but the extension of home-ownership and increase in tax liability have made them of unparalleled importance. In the mid-1980s the total cost of mortgage interest tax relief was £2,750 million and that of capital gains tax relief £2,500 million. If such cash flow measures are not regarded as appropriate measures of housing assistance rigorous analysis suggests that owner-occupiers in 1983 on average received substantial subsidies (about £170 per year) while the average local authority tenant no longer received any subsidy (Ermisch, 1984, pp. 45, 50).

Fourthly, the increased demand for private housing has not been accompanied by adequate supply-side response. The private rented sector has continued to decline. Its decline and the problems of its continued existence were even increased by the Rent Act 1957 and attempts to resuscitate it. The private speculative builder, building principally for sale, has played a major role in meeting housing needs but has not provided on the scale and at a cost which would be necessary for a fuller role. Rather than creating free market conditions government has consequently used subsidies and direct intervention to increase the role of the market sector. The most marked example of this is the right for council tenants to buy their dwellings but a wide range of mechanisms have been adopted (and consolidated in the Housing Act 1980). The evidence suggests that they widen the choice of those who can afford to buy rather than extend the opportunity to buy to new sections of the population. The problem of providing high quality housing for those on lower incomes other than through renting has remained. As owner-occupation has grown to embrace a wider range of income, age and occupational groups, the need for continuing state support for home-ownership – not just to enable people to become home-owners for the first time but to meet the costs of repairs and maintenance and to deal with crises affecting ability to pay – has become more apparent.

Fifthly, legislation and regulation have significantly changed the responsibilities of local authorities to deal with a wider range of housing issues. Legislation has introduced improvement grants (1949); regulated standards for improvement (1954); introduced area improvement policies through General Improvement Areas (1969) and Housing Action Areas (1974); encouraged local authority mortgage lending (1959); introduced a mandatory rent rebate and rent allowance system (1972) and extended the scale and supervision of this through the housing benefit system (1982); severely restricted local discretion over rents and rate fund subsidies (1972 and 1980); imposed duties in relation to homelessness (1977); increasingly regulated local authority capital programmes (under the Housing Investment Programme system after 1977); and regulated council tenants rights and conditions of tenancy (1980). Local authorities housing powers and duties have grown and they have increasingly organized and planned to provide for a comprehensive range of needs in all tenures.

Sixthly, the encouragement of the voluntary sector and the development of the fair rents system in the privately rented sector have resulted in the creation of a substantial 'fair rents sector'. In the 1980s the increase in council rents resulting from central government's use of the subsidy scheme under the Housing Act 1980 have been a final element in rationalizing rents throughout.

Regulation of the private rented sector has incorporated a fair rents system (in 1965) operated by rent officers independent of local authorities. The Housing Act 1964 and the establishment of the Housing Corporation commenced a substantial programme of state support for housing association housing as an alternative to council housing. The housing association sector has developed until it is in the 1980s a third arm more directly controlled by central government than is council housing but operating a substantial programme which had grown while net local authority expenditure had fallen. In

1982–3 net housing corporation expenditure was £680 million and gross loans and grants to housing associations £755 million. By comparison net local authority expenditure was £707 million and gross local authority capital expenditure (excluding improvement and thermal insulation grants and lending for house purchase and improvements totalling £500 million) was £1,932 million. Of this latter figure £134 million was loans and grants to housing associations (Treasury, 1984, p. 50)

A Changing Housing Service

The organization of housing in 1985 differs markedly from that of 1935 for all of these reasons. In the fifty years since 1935 council house building has moved through a number of phases from residual slum clearance (1935–9), wartime halt (1939–45), dominant high standard building role (1945–57), residual slum clearance (1957–64), a broad equity in declining output (1964–80) and a new residual role (after 1980). The new residualism after 1980 has involved the lowest ever peacetime rates of council house building. It has also involved the highest rates of council house sales and has involved an absolute decline in the size of the sector. This characteristic of expansion has given way to decline. The direct role of local authorities in housing is diminishing. But quantitative decline is only part of decline. When Sir Ernest Simon was writing almost all council housing was newly built to high standards. In 1985 some purpose-built council housing is over sixty years old. In addition councils have acquired older properties – often ill-maintained inner city dwellings. The role of improver or demolisher has involved councils in becoming slum landlords and in adjusting their stock to reduce the average standard of accommodation. Finally the change in housing standards and the fashion for high-rise and system building in the 1960s has left local authorities with a stock of dwellings of varying standards, sometimes with severe design faults and often with little consumer appeal. Such developments have significantly contributed to a loss of public support for council housing. Such a de-legitimization not only is a product of high-rise building or other changes in the stock but also has been affected by other developments which have left council housing with a more definite 'second-best' image than in the past. The financial advantages of owning rather than renting are part of this, as is the constant urging of the merits of home-ownership.

In this perspective it is important to acknowledge continuing local differences. Council housing in, say, Hackney has a very different meaning from council housing in, say, Hyndburn – both in terms of the types of dwellings it comprises and in how it compares with home-ownership in the same area. But in both areas the factors determining what is there and how it is valued are increasingly results of a greater central control and nationalization of housing policy. The high-rise building boom was orchestrated by central government subsidy and the capacity and perception of builders and building professionals (Dunleavy, 1981). Increasing co-operation between local authorities and builders over this issue or more recent partnership schemes or similar co-operation with building societies has developed in an atmosphere in which local authorities must adjust to other agencies if they are to achieve anything in housing. The consequence is policy-making in a more constrained and diversified environment. The extent to which some aspects of public housing (and especially what is built) are considerably influenced by interests and fashions operating at a national level, has changed since 1935, and has become more apparent.

This nationalization of housing policy has other dimensions. It is argued that over the 1970s and 1980s legislative and other developments have changed central government's approach to local authorities in housing, from a *laissez-faire* or bargaining approach to one of regulation and control (Malpass and Murie, 1982). The detail and complexity of controls (particularly financial controls) has increased. Indeed cuts in expenditure, the operation of the right to buy, the subsidy system (and rent and rate contributions) under the Housing Act 1980 and the rules governing the use of capital receipts from the sale of housing and housing land have combined to relegate the public housing programme to a small, residual and substantially self-financing programme.

Aspects of allocation, tenancy, rents and rebate systems all show evidence of greater nationalization of approaches. In some areas of policy central government's monitoring and supervision has become very substantial. The clearest example of this is the right to buy under the Housing Act 1980. In December 1983 the government was in contact with about 200 local authorities about aspects of their performance in implementing the right to buy. This is a much more active intervention than has generally applied in the housing area and the legal action against Norwich City Council over this same issue represents an action by central government to make an example of a responsible authority rather than a check to an authority refusing to implement legislation as in the case of Clay Cross DC following the Housing Finance Act 1972. These factors themselves and the increasing complexity and comprehensiveness of policy have increased the professionalization of housing. The reorganization of local government in 1974 substantially reduced the number of local housing authorities and the style of management adopted has reflected general changes in local authority management. Accountability to tenants or applicants for council housing has not increased and the shift of power has almost certainly been to managers at both central and local levels.

One of the most striking changes in relationships has been the shift from financial support for council house tenants as a group through general assistance subsidy, rate fund contributions and rent pooling to support for individuals through means tested rebates. The trend from collective to individual support has also involved a move from local discretion (over rents and rebates) to more closely regulated local administration of a national scheme.

Under the Social Security and Housing Benefits Act 1982, local authorities (whose rents have generally risen as a result of the way central government has carried out its powers under the Housing Act 1980) have become the local agencies for the administration of rent rebates to all tenants including those receiving supplementary benefits. The regulations covering this administration are detailed and the relationship between the local authority as landlord and the tenant is complicated by a role as social security agency. The character and nature of local authorities in housing has moved a long way from that of 1935. Exchequer subsidy, new building and slum clearance are now smaller aspects of a service concerned with homelessness and rationing access to housing, with sales of dwellings and encouragement of and assistance to owner-occupation, with repairs and maintenance of a dwelling stock of mixed quality and reputation and with the administration of means tested benefits.

The final and even the most obvious element in the nationalization of housing policy has involved the shift in balance of state expenditure and influence. Over the period since 1951 central government has increasingly exercised its housing policy through fiscal measures, through incentives to private individuals and organizations (builders, building societies and landlords) and through the more malleable vehicle of housing associations. Local authority discretion has had a decreasing role in housing and local authorities themselves have increasingly been drawn into or forced into policy innovations into which they previously were reluctant to enter.

The impact of these developments in general is that council housing has for a considerable period of time, and in association with the decline of private renting, been developing into a residual or welfare housing sector. Sir Ernest Simon's references to the quality of council housing and the council service contrast with the tarnished image and the low quality of service received by many council tenants. In 1985 commentaries are more likely to refer to a Poor Law service available on less eligible terms to households unable to fend for themselves. While the service varies enormously, new council tenants are probably more likely than at any previous period in peacetime to have experienced homelessness and also to have to accept less desirable properties especially on badly designed inner city or isolated estates. Those who are unemployed or in the least well-organized and well-paid sections of the labour force are now more concentrated in council housing. This is now less a matter for congratulation about channelling good housing to poor people but more a matter of concern over multiple and mutually reinforcing disadvantage. The potential role envisaged for council housing has not been realized.

Future Directions

The task of the next generation referred to by Sir Ernest Simon in *Municipal Progress* – 'the planning and rebuilding of the residential portions of our cities' (p. 218) – has not been achieved. The rate of modernization has been slower. Some developments have themselves come to be seen as part of the housing problem. This is most obvious in relation to high-rise non-traditional building types in both public and private sectors. But it also applies to patterns of financial and policy support which contribute to delaying inner city areas to over-consumption in some areas and under-investment in others and to an increasing association between low income and housing problems (Murie, 1983).

In 1985 the task probably seems even more formidable than in 1935. Statistics of housing stress, of disrepair and obsolescence in the housing stock and of the rate of new building and replacement indicate that an enormous task of rebuilding remains. But the coincidence of favourable circumstances in 1935 has not been reproduced in 1985. The private sector has not been able, in spite of massive support, to take hold of the task. But the less optimistic interpretation of the last fifty years is that the barriers preventing our city fathers from accomplishing the task are considerable. The failures of the interwar years were not signs of temporary problems but were the result of real conflicts of interest and an unwillingness to devote resources to housing those experiencing the greatest problems.

Even if economic, public expenditure and political constraints gave the go-ahead the high-rise experience has left doubts about the capacity of the construction industry, the local authority or the policy machine generally to deliver the rebuilding required. In addition, the development of housing tenure and of interests around different tenures have created new barriers. Many commentators would agree that without some rationalization of the structure of housing finance the capacity to deal with outstanding problems is severely diminished. But the political barriers to such a rationalization are immense.

These factors now complicate and reduce the capacity to deal with the housing of the poor – to bridge the gap between the costs of adequate housing and levels of income. One view of the pattern of housing development since the war is that it has represented a response to the demands of organized labour or those sectors of the labour force who have political and economic importance. The residual or marginal or 'undeserving' poor have never sufficiently been a central target for improved housing. The poorest and those with least bargaining power have consequently continued to benefit least. These are not accidents of policy but consistent consequences of broader patterns of economic and political power. If the failure of the fifteen years before or the fifty years after 1935 are not to be reproduced subsequently the problems involved are more deep-seated than Simon suggested. What is involved in the progress that Simon envisaged is breaking a pattern in which the poorest have often not benefited as much as those who are better off. This applies to the local delivery of services as well as decisions about finance and resources. The debate on housing policy in the 1980s is increasingly about rights. For local authorities to be able to provide rights to a home, rights to buy, rights to sell, rights to mobility and choice, and rights to a clear quality of service, fundamental changes in financial arrangements and resource expenditure and probably also in the organization and control of the housebuilding process, housing costs and housing tenure would be necessary. It may be that changes in tenure would even mean an end to the landlord role which local authorities have developed over the history of council housing. The role of developer may prove a more important and continuing one and it may be that the future contribution of local authorities will be in building and improvement and in facilitating exchange of dwellings rather than in managing and owning a substantial stock. Without some such substantial changes it is unlikely that Simon's vision will be realized. The present direction of policy abandons the collective for the individual and the local for the national. It provides a recipe for residualization and enhanced inequalities in housing. The alternatives are not restricted to the converse of this, or to a reversion to some historical formula, or even to neutrality between existing tenures. There is scope for the development of local initiatives and responses to local housing problems which maintain and enhance individual rights *and* acknowledge areas where collective provision works for the individual. The challenge for the next generation is to redirect the housing debate rather than be constricted by its momentum.

10

Local Government and Social Policy

NICHOLAS DEAKIN

> When a great, and urgent, and difficult national problem has to be dealt with the first necessity is to choose carefully the men to do the work and to give them whatever powers and resources they may need to accomplish it. It is well that this first step has now been taken. Yet the analogy of the new departure with the entrusting of Poor Law administration to the 'three Bashaws' was too close to allow us to look forward without misgiving.
>
> Gilbert Slater in *A Century of Municipal Progress*, p. 369

Gilbert Slater's chapter was written shortly after Neville Chamberlain's redistribution of Poor Law functions had broken down under the impact of mass unemployment, which (Slater observed) might be 'the rock on which our present social order will split and capsize' (p. 368). New legislation, the Unemployment Act 1934, provoked much head-shaking about the authoritarian style in which the powers of a new central government agency, to which many of the functions of local government in this field had been transferred, were being exercised. But beyond these immediate issues lurked other problems, still only dimly glimpsed. 'Looking towards the future', says Slater, 'we see that Malthus has triumphed. Births numbering only 580,413 in 1933 against 734,788 persons married makes fears of an excess number of births ridiculous. They have given place to new fears of deterioration in the average innate quality of children born, and of racial suicide. These fears will create future issues with regard to public assistance, which will doubtless continue to be fertile of social conflict' (p. 369).

Time has not dealt kindly with this kind of prophecy. When Slater wrote, concern about population composition was widespread. Prime Ministers could write, as Ramsay MacDonald did, that 'that old ruffian Malthus' had turned out to be correct. Richard Titmuss, in the first of his many distinguished contributions to social policy, *Poverty and Population*, concluded that:

> From the more restricted viewpoint of the intelligence level in specific communities and areas with declining populations, it does not appear unreasonable to postulate that intensive migration of the best types of individuals drawn mainly from the young age groups will have detrimental effects. The possibility, for instance, of marriage for those who remain will be confined. It is not unlikely, as experience has shown, that greatly depopulated communities with an unfavourable environment will tend to develop into 'clusters of subnormals'. (Titmuss, 1938, p. 47)

Since 1938 eugenics as social engineering has been discredited by the experience of Nazi rule in Germany; and the contributions of the Eugenics Society, the principal source of propaganda on this issue, to discussions of postwar social policy were confined to an eccentric footnote to the debate on immigration control after the Second World War. When one senior politician, Sir Keith Joseph, attempted to revive some of these issues while in Opposition in 1975, his speech was widely considered to have forfeited him the chance of leading his party. But one element in the prophecy has proved accurate enough: public assistance, in its latter-day forms, has indeed been 'fertile of social conflicts' and local government remains at the centre of the debate.

The Relief of Poverty

The temptation, Dr Slater's gloomy analysis notwithstanding, is to see the 1930s as a (rather pallid) golden age for local government. It is certainly possible to identify in the portrait provided by *Municipal Progress* a number of grounds for such a judgement: the relative independence from central government, the persistence of the traditions of innovation from below and the growth of municipal enterprise. But although local government's writ still ran in areas like health and public assistance which were subsequently to be lost to it, in most major areas of social policy, the powers and functions of local authorities were, as a series of reports produced by Political and Economic Planning (PEP) on the social services in the course of the 1930s make clear (Pinder, 1982), a pale shadow of what they were subsequently to become.

Events during and immediately after the Second World War changed all that. The conventional view is that local government shed its trading responsibilities and substituted a major role within the welfare state. In fact, the pattern is not quite so tidy. The Beveridge Report on Social Insurance and Allied Services (1942) is rightly seen as the keystone of the welfare state structure. But it defined only a very limited role for local government. Beveridge brusquely dismissed any notion of entrusting local authorities with the new social security function:

> It is neither desirable for the strength of local democracy that local authorities should have to administer a service without discretion and according to detailed rules, nor compatible with the national minimum and a National Plan of Social Security that people should be given assistance according to a scale of conditions which vary from place to place. (Beveridge, 1942, p. 67)

Local government was to be left with a number of residual functions 'under a special social welfare committee concerned with or co-ordinating such cash-aid functions as rent rebates or cheap or free school meals to which a single means test would be applied' (ibid., pp. 68). Moreover, Beveridge's proposals were based on the premise that three other further major innovations would occur: the containment of unemployment to an acceptable level; the provision of family allowances; and the creation of a national health service. None of these was to fall within local government's sphere. Coming from Beveridge, this emphasis may not have been surprising. As an academic and former permanent secretary, he had limited experience of local government and a distaste for political conflict – 'the greatest hope of dealing with unemployment', he once wrote, 'lies in treating it not as a political issue but as a technical problem' (Madge, 1943, p. 9).

Such an approach was consistent with the general enthusiasm for national planning as a tried and tested technical activity which would adapt the techniques that had won the war to 'winning the peace'.

Attlee's government inherited this approach from the Coalition (in which all its major figures but one had served). Centralized planning would be conducted at the national level, to which would be added the distinctive emphasis on the goal of equality, sanctified through common experience in the war years and the applications of the tools of social science to achieving efficiency. In this way, the promised reconciliation between the objectives of democracy and efficiency would be brought about. As Herbert Morrison, who was responsible as Lord President for the co-ordination of Labour government policy, put it:

> Britain was the first great nation to attempt to combine large-scale social and economic planning with a full measure of individual rights and liberty. Planning as it is taking place in this country under our eyes is something new and constructively revolutionary. (Morrison in Donoughue and Jones, 1973, p. 353)

A valid alternative tradition existed for locating responsibility at local level – the Webbs had written in *A Constitution for the Socialist Commonwealth of Great Britain* that 'there seems practically no limit to be assigned to the number and range of the industries and services which might advantageously be undertaken by local authorities' (Webb and Webb, 1920, p. 236). Local government had a potentially powerful spokesman in the inner Cabinet in the shape of Morrison himself, architect of the London County Council's prewar success. But in the construction of the remainder of the welfare state edifice the key responsibilities for framing social policy were entrusted to central government departments and for much of its implementation to non-elected agencies and boards.

Most conspicuously, the health service was lost to local government when Bevan defeated Morrison in Cabinet. When, in the words of the National Assistance Act 1948, it was enacted that 'the existing Poor Law shall cease to have effect', the bulk of the residual responsibilities went to a National Assistance Board, as Beveridge had recommended. Furthermore, the major visible expression of integrated physical, economic and social planning, the new towns, were placed wholly outside the local government system under the direction of Lewis Silkin's Ministry of Town and Country Planning, and managed by appointed development corporations.

As part of the postwar reforms, a very rapid expansion of local government functions in social welfare did duly take place, but under close central government direction. The first major piece of legislation, R. A. Butler's Education Act 1944, contained in section 1 a clear statement that the local education authorities would function under the 'control and direction' of the minister. From the organizational perspective, education could legitimately be seen, in a much contested phrase, as a 'national service locally administered'. Other services followed the same pattern. The Children Act 1948, which created the local authority children's services, provided that local authorities should exercise their functions 'under the general guidance of the Secretary of State – provided through the scrutiny of an inspectorate whose job was to secure efficiency and ensure that the detailed regulations issued by the Home Office were being observed. Housing, which narrowly avoided becoming a national function, was subject to particularly close financial control during a period of acute scarcity of resources. The much expanded functions of planning which, despite the wartime enthusiasm for central planning agencies, were eventually entrusted to local government under the Town and Country Planning Act 1947, were also made subject to ministerial scrutiny and review (Cherry, 1982).

All these functions, old and new, were exercised under close ministerial direction, established in legislation, supplemented by statutory instruments and, in the case of disobedience, subject to strong default powers. In several cases the appointment of chief officers was made subject to departmental veto. Strict application of the *ultra vires* rule confined authorities to the tasks as defined in statute.

The local authorities that were to administer these vastly expanded functions were in all essential respects those of the prewar period. Furthermore, the government showed little or no interest in structural reform. In the case of health Bevan, as responsible minister, had inherited a coalition government White Paper; but in contrast to his response on the health service, Bevan showed little public interest in local government on his own account. In Robson's harsh judgement, he displayed 'no trace of any zeal for radical reform of local government nor any strong belief in its virtues' (Robson, 1966, p. 91).

In consequence, the new services had to be set up and administered within the existing system, and their organization duly followed the traditional model. As Glennerster describes it:

> The pattern of a specialist or professional chief officer reporting to a committee of lay people was well established before generalist local authorities came into being. This pattern became firmly established as more powers and duties were added to local government's load in the twentieth century. The result of this accretion of specialist committees was what came to be called the 'traditional' system of local government: dominant, vertical or service based committees serviced by chief officers with distinct professional training, and relatively weak 'horizontal' committees such as Finance and Establishments. (Glennerster, 1983, p. 38)

The separate departmental structures created and maintained as a result of the welfare state reforms were staffed and run by specialists; the absence of a national cadre of local government staff and the low esteem with which their generalist administrators within the system were regarded reinforced these powerful and exclusive cultures. The extension of the ethos of functional separation to the new social policy departments created in the immediate postwar period was (as Stewart shows in Chapter 5) entrenched by the rapid professionalization of these services. Planning, housing and social work: each in its turn passed rapidly through the same process of establishing criteria for qualifications, credentials and training courses. Staff scrambled for the security that a formal qualification conferred. Griffith, in his review of the relationship between central and local government conducted in the 1960s, comments that:

> The professionalism of local government officers is the greatest single force which enables local authorities to carry out, with much efficiency, the considerable tasks entrusted to them. And pride in a profession is a better insurance of first class performance than more material incentives. (Griffith, 1966, p. 534)

Later experience casts some doubt on the validity of that judgement; nevertheless, the presumption was that national policy implemented by local authorities as agencies of central government departments would knit together in a sufficiently coherent and efficient whole to ensure that additional measures for the relief of poverty would not be required.

By the late 1950s the new pattern of provision was firmly established. Some degree of relaxation of direct control from the centre had taken place: the system of general grant introduced in 1958 was matched by a declaration of intent by central government. Control would be limited to the 'key points' of a service and 'though for this purpose basic standards may be prescribed, the maximum local discretion will be given in the method of providing the service' (Griffith, 1966, p. 504). Moreover, the achievements of the new system of welfare were considerable. Beveridge's preconditions of full employment and the creation of an effective health service had been satisfied. The process of eliminating gross disadvantage and extreme regional disparities appeared to be underway. The reformed education system was providing some evidence of making new openings for talented children. The worst slums had been demolished and the programme of municipal house construction vastly expanded. Macmillan's coup of achieving the construction of 300,000 houses became one of the most frequently celebrated achievements of the Conservative government of 1951–5.

More striking in many ways than any formal achievements had been the general acceptance of the view of the welfare state that T. H. Marshall encapsulates as 'the *tableau vivant* of the good society and a permanent, praiseworthy and admirable feature of twentieth century life' (Marshall, 1970, p. 12). The Conservatives, on assuming power, had expressed concern about the cost of services; but the principle of state provision of welfare in which local government performed an essential (if subordinate) role went virtually unchallenged. As clear-sighted an observer as C. A. R. Crosland could write optimistically in *The Future of Socialism* (1956) that the basic problems of poverty were on their way to being solved and it would be in the area of specialized services to cope with new needs that expansion would be required.

Yet a number of crucial questions remained unresolved. Richard Titmuss summarized many of them in his Fabian pamphlet, *The Irresponsible Society* (first published in 1959). 'There is little evidence from the history of the last ten years', he said, 'that society is any nearer to the solution of these problems of dependent poverty, inequality and unfreedom. By any objective criteria of wealth and opportunity it is in some respects further away' (Titmuss, 1963, p. 217). By way of illustration he cited local authority expenditure on welfare services, which had barely increased in real terms over a decade. This failure to make significant progress he attributed to 'the shifting concentrations of economic power in a more prosperous society' and the development of a system of fiscal and occupational benefits that provided for the better off, leaving a residual 'welfare state for the working class' – which was increasingly perceived as a burden on public resources. He concluded:

> To gain affluence then does not mean that we should abandon the quest for equality. In some senses at least the quest becomes harder to undertake as the cruder injustices of yesteryear are reduced and blurred. But new forms and manifestations of social injustice take their place. (Titmuss, p. 243)

One such manifestation which had emerged from obscurity over the course of the decade was the issue which was then generally described as 'colonial immigration'. The decision of Sir Anthony Eden's government in 1955 not to proceed with legislation to control this immigration allowed numbers to build up to substantial levels in the second half of the decade. Local authorities in areas of immigration were presented with a series of new issues on which they were bereft of the customary guidance from central government, once the Colonial Office had divested itself of its residual responsibilities for welfare. The appearance of a visible minority in several major conurbations revived fears about what one politician chose to describe as an 'alien wedge' – a phrase reminiscent of eugenic anxieties about pauper immigration earlier in the century. After the events of 1958 in Notting Dale and Nottingham, when Blacks were victims of what the press chose to describe as 'race riots', additional anxiety developed about public order. But the anxieties were still not sufficient to persuade government that any special measures to provide for housing, education or social services for new immigrants were necessary. Forceful representations by local authorities were met with assurances that equality before the law and right of equal access to welfare services would suffice (E. J. B. Rose *et al.*, 1969).

By 1961, under heavy constituency pressure, the Conservative government decided that the value of common British citizenship with its associated right of free entry in a changing Commonwealth no longer outweighed the political risk and introduced legislation to control immigration. As a political manoeuvre, the expedient proved brilliantly successful. The Labour Party, which had opposed the legislation as being hasty, ill-conceived and discriminatory found itself pilloried by its opponents as the party of the open door. But legislation was not followed by the assumption of any responsibilities beyond the keeping of the gate – a progressively harsher system of control vested in the Home Office. An advisory committee (the Commonwealth Immigrants Advisory Committee) was the furthest that Whitehall was prepared to go. Even when the issue forced its way on to the centre of the political stage in 1964, with the defeat of Labour's Foreign Secretary-elect at Smethwick, local authorities' anxieties about service provision in multiracial areas were still being brushed aside. In 1965 a White Paper (Home Office, 1965) of almost incredible blandness was issued: the established services of the welfare state would cater for all foreseeable needs; special provision was neither expedient nor necessary; voluntary efforts with suitable adaptation would suffice. In the broader sense, the episode perfectly illustrates the limitations of the welfare state machinery and the inability of departmental structures, centrally and locally, to adapt to new needs.

Towards Professional Management

By the time this White Paper was issued, Labour was back in power, with an agenda elaborated in the long years in Opposition. The main emphasis in social policy was to be on the modernization of the institutions of the state and the professions that ran them. The programme of reform was to be achieved chiefly by classic techniques of public inquiry informed by the professional insights of the newer professions – economists, planners and social scientists.

The first exotic flowering of this approach was George Brown's *National Plan*. This was to be blown away in the economic crisis that descended all too rapidly on the government. But the reinstatement of central planning as a major emphasis in policy which the plan symbolized had a more substantial impact through the creation of a new structure of national and regional economic planning.

The *locus classicus*, however, was the inquiry into the Civil Service chaired by Lord Fulton (Fulton, 1968), with its withering opening denunciation of the amateurism of British government and its civil servants. The recipe for reform that Fulton produced was to be of general application: the central importance of the possession of general management skills; the definition of specific sets of professional expertises (of which 'social administration' was to be one);

reform of the system of recruitment and training; identification of priorities for policy by the creation of special free-standing planning units supported by the provision of analytical capacity and intelligence; and the sovereign importance of the co-ordination of policies in related areas.

A succession of similar inquiries distilled similar messages. As Brian Abel-Smith subsequently put it, 'reorganisation was the key to comprehensive social planning and the assertion of rational priorities' (Glennerster, 1983, p. 14). At the request of the local authority associations, the Conservative government had in its last months launched two separate inquiries into local authority management and manpower: both reported in 1967 (Maud; Mallaby). But before they did so, Richard Crossman took advantage of a convenient speaking engagement, and the absence of his formidable permanent secretary, Dame Evelyn Sharp, to throw the whole system into the melting-pot by setting up a Royal Commission (Redcliffe-Maud, 1969), the detailed origins and consequences of which are unravelled by Alexander (Chapter 3).

Meanwhile, perhaps most centrally of all, personal social services passed under review. The Seebohm Committee on Local Authority and Allied Personal Social Services, appointed in 1965, was based on the assumption that services for the family as they had been distributed after the postwar settlement between children's departments, welfare departments and social work in housing and education, were too widely dispersed. The key problems identified were those of co-ordination and access, their manifestation and failure to reach those in most need.

The conclusions of these two inquiries left an indelible mark on the organization and style of operation of local authority services. In the case of Seebohm, the new large social service departments created as a result of the reform system introduced and institutionalized a new approach by local authorities to social problems. As the committee themselves put it, the creation of the new departments was to:

> ensure a more co-ordinated and comprehensive approach to the problems of individuals and families in the community in which they live. It should be more effective in detecting need and encouraging people to seek help; it should attract more resources and use them more efficiently and it will be possible to plan more systematically for the future. (Seebohm, 1968, p. 44)

One of the subsidiary themes that Seebohm chose to address was that of poverty, identified in this instance in terms of 'areas of special need', defined as 'localities which have a profusion of pressing social problems, offer only a dismal and squalid physical environment, are inadequately served by social services, and are considered to justify special attention and a generous allocation of resources' (ibid., p. 149). This identification stemmed in part from a growing recognition that poverty could not be treated as a residual, but required action in its own right.

It is customary to refer to poverty being 'rediscovered' at this time after a period of complacency about the effects of welfare state reforms. In fact, it is difficult to identify a period in the postwar years when there was not some recognition of its persistence, although the publication in 1965 of *The Poor and the Poorest* is generally reckoned to mark a turning-point (Abel-Smith and Townsend, 1965).

The pedigree of special action to cope with poverty on an area basis is clear enough. The Milner-Holland Committee on London Housing (1965) proposed areas of special control. This proposal was taken up and extended by the National Committee for Commonwealth Immigrants (NCCI) as 'areas of special housing need' which led in turn to the introduction of new powers in housing legislation to declare General Improvement Areas. Meanwhile, in the related field of race relations, pressure from the NCCI had led to the introduction of powers in the Local Government Act 1966 to pay for staff employed on services related to the presence of Commonwealth immigrants (section 11). The Plowden Committee, reporting in 1967, proposed an equivalent intervention in urban primary schools, under the rubric of 'positive discrimination'; Educational Priority Areas (EPAs) were to be created, special projects set up and additional resources committed to schools within EPAs (Thornbury, 1978, p. 22).

The various streams flow together in the action taken by the Prime Minister in the aftermath of Enoch Powell's speech on immigration in 1968, when 'like the ancient Roman' he 'seemed to see the River Tiber flow with much blood'. Although the race issue was the trigger in the creation of the Urban Programme and several would-be participating local authorities imagined at the outset that they were engaged on a competition to demonstrate the existence of need through the presence of Blacks, the Urban Programme swiftly moved away from a focus on race. Once embodied in legislation, the

Local Government Grants (Social Needs) Act 1969, it became a general fund from which central government supported a substantial number of projects nominated by local authorities in the statutory and voluntary sector to the tune of 75 per cent of expenditure incurred. Thirty-four local authorities were invited to participate initially on the basis that they contained identifiable small areas of concentrated social problems, though these were defined from the outset in very wide terms.

The Urban Programme was to develop and change as it passed from improvisation to institution, and eventually emerged not as a strategy but as an expenditure mechanism – a device for spending a limited quantity of additional funds on terms approved by central government 'in areas deemed by local authorities to have special social needs' or 'concentrations of urban deprivation' (Edwards, in Higgins, 1983, p. 60). Local authorities became accustomed to the system of bidding and treated the programme as a useful supplement to their funds, or as Danegeld for importunate voluntary organizations. Government sought to assert some control over the content of the programme by stressing various themes (under-fives in the early period) but ended by spreading the not very plentiful jam so thinly as to frustrate this objective.

The programme did not disturb the fundamental assumption that the additional problems that had emerged as a result of what was still termed 'urban deprivation' could best be dealt with by the main welfare state services, their priorities modified if necessary in directions to be determined by central government departments, as Crosland and Crossman sought to do in education and housing respectively in the lifetime of the first Wilson government. Behind this approach, in turn, lay the basic assumption that had underpinned the whole range of activity in social policy in the postwar period: that the economy, managed with the help of Keynesian macroeconomic devices, could be relied upon to deliver the resources necessary to deal with the new problems which were constantly being exposed. By the end of the 1960s that assumption was wearing a little thin: the revival of a full range of planning instruments, so successful in the late 1940s, had not succeeded in correcting the British economy's well-established propensity to descend into periodic crisis. Moreover, regional policies, even when reinforced by new institutions of regional economic planning, had manifestly failed to solve the problems of the depressed areas. In many ways more ominously still, deterioration was beginning to be observable in the economic well-being of some of the major conurbations that had previously provided the focal points for growth (see Dawson, Chapter 2).

In 1969 a working party of the Labour Party's National Executive Committee produced a comprehensive review of social policy which contained and consolidated a series of arguments for a new co-ordinated approach to the planning of welfare. Social policy should not be seen only in terms of developing the postwar welfare state, or in isolation from the industrial system, the working party argued (p. 14). As a major priority, the working party wanted to see a Priority Areas Programme embracing all the social services:

> In the identification of needs and of three criteria by which to select areas of positive discrimination, the role of central government in relation to local authorities must be developed far more. It should prod them to widen their activities and raise their standards. It should obtain the information from local authorities and help them to evaluate their own needs by comparison. There is no machinery for them to do this at present. Instead of merely stating what might be national minimum standards of provision, central government departments could begin to define standards by reference to what already exists in the better areas. Discrimination on the scale required demands a much stronger role for central government. (Labour Party, 1969, p. 73)

Such a role would require the creation of a central planning unit – designated the Office of Social Planning, and covering the whole field of social policy (p. 84). But before any systematic action could be taken, and while the reform of local government based on the Royal Commission Report had not yet been translated into legislation, the Labour government was unexpectedly defeated at the general election of 1970, and the Conservatives returned to power.

The Sea Change

The new Heath administration made it clear from the outset that the government would be committed to improving managerial efficiency in government. A White Paper on the machinery of government, the setting up of two large departments of state, the Department of the Environment (DoE) and the Department of Health and Social Security (DHSS, this latter including the children's department, formerly located in the Home Office), and the creation of a Central Policy Review Staff (CPRS) were early indications of this commitment. If the reform of local government that eventually took shape was a great deal more cautious than the proposals in the Royal Commission report it was nevertheless sufficiently comprehensive to provide the basis for a series of new departures in social policy.

The reformed local government system rested on a concept of corporate rationality with a stress on the creation of a strong central focus. As Stewart shows (Chapter 5), it presupposed a political leader, a new forum for determination of policy priorities in the policy and resources committee, and a chief executive officer who would be serviced by a central analytical capacity. Together, these innovations were intended to enable authorities to become more responsive to the needs of their community and to present a collective front to central government. The message was especially welcome in the new upper-tier authorities which had assumed the strategic planning role. One use to which this powerful new set of tools was put was to identifying common social policy goals: as was the style of the moment, the DoE commissioned management consultants to undertake studies that would identify 'urban guidelines' for the new authorities to follow in addressing social issues.

Something of the sort had already been attempted in London, where reform had taken place a decade earlier. The Greater London Council, born of the Herbert Commission inquiry, had been presented with the task of producing a Development Plan (GLDP) which would identify strategic planning goals for the whole conurbation, not constrained by the boundaries of local authority service provision or the powers of those authorities. The authors of the GLDP set out to provide as part of that process a credible set of social objectives: the popular statement *Tomorrow's London* that accompanied the plan reflects this concern:

> It may seem that the preoccupation of the Development Plan is too much with the evils of the past – that we think we can build a new world simply by dealing, one by one, with the known maladies of the old. But this is not how we see it. Our new London is geared to new demands, new forms of life, new technologies and of new uses

> for our human faculties. The Plan is a beginning – and not the end. (GLC, 1969, p. 79)

With a plan covering a canvas as broad as this it would be pardonable to wonder who could undertake such a wide-ranging task. *Tomorrow's London* offers one candidate: the planner, who is

> concerned with the physical environment, but he is equally deeply involved with the efficient functioning of the economy, the growth of communities and the correct use of scarce resources for which there is competition from many directions. This means the planner becomes a listener, a researcher as well as a decision maker. The office of the planner is to gauge trends in the economy and in society, to pinpoint shortcomings and sources of frictions, and to make proposals for solving the immediate problems of the community in which he works . . . . In these processes the planner cannot work in isolation. (ibid., p. 44)

The planning profession had indeed gone through a series of evolutions; among them the fashion for a 'systems' approach and the related requirements of numeracy and acquisition of new techniques of analysis. By the beginning of the 1970s planners were confident enough to undertake such a role. David Eversley, an economic historian who spent three turbulent years as the GLC's Chief Planner (Strategy), reflected after his departure on the implications of the successes and failures of the new planning as he defined them. 'It is not a question', Eversley concluded, 'of whether the planner wishes to play the role of arbitrator, whether he is content or loath to accept his social role: he *faces* the task, and he is unprepared for it' (Eversley, 1973, p. 97). In this process, social planning must be a major part of his responsibilities and the tools of social science (social surveys and demographic analysis) will be an essential piece of equipment. Yet, asked Eversley, 'how is he to do his work within the confining limits of the existing structure of local government? This is the question the Maud Commission never answered' (ibid., p. 142).

Nevertheless, when the new authorities set up under the 1972 Act outside London came into existence, a systematic attempt was made to introduce these new modes of planning. The key stimulant, as ever, was the approach adopted by central government: different departments approached the new situation in different styles; but a new feature in the situation was the attempt to achieve some form of co-ordination. The CPRS, immediate offspring of the new movement for rationality in government and remote descendent of the Attlee Labour government's co-ordinating Lord President's Office, produced in 1975 the *Joint Framework for Social Policy*, an attempt to exploit the new corporate structures in government as a whole and create a new awareness of the need to define priorities for social policy – identified in terms other than those imposed by the historical accident of the allocation of departmental responsibilities in Whitehall. The CPRS's willingness to include local authorities in this process was part of a broader definition of the issues; but neither across Whitehall, nor between Whitehall and town hall did this initiative leave any permanent mark.

However, social planning did succeed in establishing itself as a concept within the new large Whitehall departments. They, too, had their visitation from management consultants; the DHSS had had to digest not only the full range of recommendations in the Seebohm Report and the new departments set up in the new local authorities, but the equivalent major reforms in the health service and development in consequence of a whole new network of relationships. In 1976 the DHSS produced a document intended to provide the framework for long-term planning in this new environment. The Secretary of State, Barbara Castle, in her introduction to the consultative document, described it as:

> a major new approach to planning. One of the biggest challenges to effective democratic government is how to reconcile two potentially conflicting aims: central government must be able to establish and promote certain essential national priorities while the local agencies of government should have the maximum scope for making their own local choices in the light of their own local needs. This document seeks to turn planning into a co-operative enterprise: a process in which guidelines from the centre are related to – and influenced by – the experience of those who have to apply them in local circumstances. (Castle, 1976, quoted in Glennerster, 1983, p. 11)

Less ambitious, the Department of Education and Science (DES) confined itself to a White Paper issued in 1972 under the title *A Framework for Expansion* (not, perhaps, Margaret Thatcher's finest hour as Secretary of State). But the DoE, having temporarily absorbed the Ministry of Transport, launched an ambitious series of new initiatives across the whole range of its relationships with local government – consolidation of area policy in housing, as General Improvement Areas were superseded by Housing Action Areas in the 1974 Act: initiation of additional work on urban deprivation through the commissioning of three inner areas studies, also from the inevitable consultants. In 1974, on the return of Labour to power, the department briefly acquired a Minister for Urban Policy: although this experiment was swiftly terminated, the department was sufficiently impressed with the need for a further new initiative in this field to embark on another series of schemes designed to try out the potential for area management on a small geographical scale.

However, the main focus of research and action on poverty within central government was still at the Home Office. Having established the Urban Programme, the Home Office had embarked in parallel with it on an experiment of an altogether more far-reaching character, the Community Development Project (CDP). The origins of this programme lay in the perceptions of a single individual, a senior official at the Home Office, Derek Morell. He argued that one of the major problems of deprived urban communities was the collapse of morale, and that this might be met by the development of resources within the community and attempting to ensure more effective delivery of statutory services, anticipating rather than responding to need.

Much of this thinking was closely parallel with that of the Seebohm Committee; it found a responsive echo in the Cabinet's Social Services Committee, where Morell described it (according to Richard Crossman's diary) as 'a second revolution in the welfare state' (Higgins, 1983, p. 18). But Morell's early death undercut the initiative at an early stage; and discontinuities in the senior management of the project at the Home Office delayed it still further. The concept was to establish specially recruited action and research teams in parallel, based in local authorities with the resources provided from the Home Office. Negotiations moved slowly, although one of the participating authorities was Coventry, where the Town Clerk (who had been a member of the Seebohm Committee) was a convinced believer. Once the teams (twelve in all) had been established frictions speedily arose: between action and research, between teams and local authorities, and between teams and the Home Office at the centre.

Partly as a result of this process, the CDP became 'radicalized' and the members of action teams moved quite rapidly from Morell's original concept of attempting a form of community development to one in which the causes of disadvantage were seen as structural and lying outside the areas in which the teams were operating. The immediate implications for the teams were that the focus of activity moved to critical analysis of government policy: this led to the creation of a central intelligence unit to discuss and help disseminate the findings.

Hence the paradoxical situation arose of a central government department sponsoring and publicizing a sequence of critical analyses, grounded in a neo-Marxist analysis of the causes of urban deprivation, culminating in one study, *Gilding the Ghetto* (Community Development Project, 1977), which was sharply critical of the project itself and of the motives of government in setting it up ('Managing the Crises'). This could hardly continue: for the local authorities concerned there was little merit in continuing to be involved in a project so far divorced from their original objectives – their local activities now caricatured by critics as advocating 'revolution on the rates'. Offered a loophole for withdrawal, several authorities took immediate advantage of it: the remaining projects were allowed to run down. Answering questions in the House, Brynmor John, as Home Office minister responsible, concluded coolly that it had had 'a negative result' and taught the government 'what not to do' (Higgins, 1983, p. 43).

This episode, tragi-comic in its detail and limited in terms of resources expended, had some far-reaching consequences. The question of urban poverty and its relief became an issue of central significance as the 1970s developed. As Dearlove argued in his *The Reorganisation of British Local Government* (1979) it was in a sense a test case:

> because the dominant ideology surrounding the welfare state promised the solution of these problems, their rediscovery poses a demand on the system that cannot easily be dismissed as beyond the proper scope of public concern and action. This poses yet another problem, however, in that the demand cannot be fully met within the confines of the established system. (p. 238)

Central government's reaction to the failure to find solutions was essentially one of impatience: the problems existed because the machinery of the welfare state was insufficiently flexible or imaginatively administered at the local level to address these problems effectively (this was the nub of Morell's original analysis). Consequently, they could be resolved only if a fresh approach were adopted; and only central government was in a position to take the analytical overview required. But two drawbacks vitiated much of the work done in the early 1970s at the centre – the unwillingness of central government departments to accept the complexity of the process on the ground, or to commit resources on a scale sufficient to make any substantial impact on them.

Local government found itself in an uncomfortable position. The additional resources were welcome; but the direct intervention of central government in local issues that accompanied them was not. The smaller sources of funding, like section 11 of the 1966 Act or the Urban Programme, could be accepted as a marginal increment to resources: but schemes like CDP, which challenged policy assumptions, posed more serious problems. The perceptions of local government staff investigated by the Institute of Local Government studies (ILGS) show them very reluctant to move outside what had already become the traditional framework of the welfare state. The researchers concluded:

> Our study has shown, not surprisingly, that there is no generally accepted definition of urban deprivation amongst those concerned with administering the local urban system. Many officers had not really considered the question previously. (Stewart *et al.*, 1974, p. 38)

Many of the new departures in the field raised in an acute form the question of co-ordination – the golden word of the reformed system, so much more easily uttered than achieved. Co-ordination in the sphere of some central departments, as in the case of joint funding developed by the DHSS across health and social services, was difficult enough to secure: the experience of the abortive joint approach to social policies (CPRS, 1975) showed how impossibly difficult it was in the social policy area to achieve any degree of co-ordination when there was conflict over objectives and the allocation of resources. Local authorities, riven by rivalries artifically accentuated in the new two-tier system, found the task of developing joint approaches between local authorities impossible. Even within authorities the powerful service department baronies survived: the GLC's poverty initiative confronted both these difficulties and ran into the sands in consequence. Even protagonists of co-ordination 'from the ground up' – the mini-town halls of the Golborne experiment in community government in North Kensington – found that the conflicts that arose between area and service perspectives were impossible to resolve in practice.

Finally, the experimental poverty initiatives confirmed suspicion on the left of the state's intentions in making these interventions. Christine Cockburn's briefly influential analysis in *The Local State* (1977) portrayed local government as using additional resources generated in this way to perpetuate local rule by officers. Such rule, she argued, was exercised in the interests of capital, from whom the corporate management techniques employed had originally been learnt. A variant of the same theme developed by the CDP teams in the last stage of their work saw the role of local government (as in Coventry) as being to serve the interests of multinational companies, who exercise the real power over the economies, not just of the inner cities, but of urban Britain generally.

New Horizons

By the mid-1970s, preoccupation with economic issues had superseded concern for urban poverty at both local and national levels. Furthermore, the concept of a social policy dividend generated by the steady growth of the economy and available to be drawn upon whenever new needs appeared abruptly ceased to have any credibility, even in the modified form in which the Labour Party NEC's working party had restated it. The effects of the major economic crisis of 1976, the descent of the International Monetary Fund and the imposition of cuts in public spending are still clearly visible as this essay is written. 'The party's over', the Secretary of State for the Environment, Anthony Crosland, explained; and so indeed it proved, even for those for whom it had never begun. Drastic cuts in capital programmes and reversal of ambitious long-term plans for expansion of services, in some cases still barely put together, began a process which still continues.

Although the Labour government's public expenditure cuts provoked protest by local authorities as well as from pressure groups and local voluntary bodies, the impact of those protests was negligible. Disillusionment with the reforms of the first half of the decade had already set in, both among the public at large and politicians at Westminster. The performance of the larger authorities, especially the new upper-tier authorities and the GLC, was a particular cause of dissatisfaction, accentuated in the case of London by conflict between the tiers which frequently obstructed or obscured the processes of social planning. The new devices of corporate planning and programme budgeting had signally failed to deliver the goods during a period of expansion, and were now switched to a new mode in which they were diverted to the task of identifying and helping to deliver cuts, not increments (Deakin, 1982). The large-scale planning of the 1970s had also been beached by the turn of the economic tide: the GLDP, despite the immense effort that had gone into its production and the protracted inquiry conducted by Sir Frank Layfield, had proved to be a damp squib and its impact on social policy negligible.

In a succession of social policy areas, public disillusionment with the performance of local government was only too plain. A series of battles over major environmental schemes had ended with unequivocal defeats for local government. Covent Garden was the most public and painful of these and became symbolic of successful public resistance to planners' unrealistic economic and social objectives. Local authority housing suffered similarly, a process much accelerated by the appalling consequences of the adoption of systems building techniques and the visible destruction wrought on inner city communities by wholesale clearance schemes. Reg Freeson became the first of several housing ministers to claim that he had 'pensioned off the bulldozer'. Public participation, which was to have reconciled the public to the planning process, became instead yet another instrument by which it could be defeated. Younger professionals deserted local government to promote 'community based' planning as an alternative approach. A proliferation of new local organizations claiming (not always convincingly) to be acting for – or in extreme cases actually to be – local working-class communities, some of them equipped with Urban Programme grants, fought local authorities at inquiries and public meetings with their own weapons and more often than not worsted them. Within the local government professions themselves – Griffith's bulwarks – the questioning grew rapidly: planning, architecture, social work, all bred their own internal oppositions, publishing in samizdat or duplicated newsletters or eventually, as the grants grew larger, in their own journals.

Attempts to generate new initiatives in the field of poverty fell foul of widespread suspicion about both their effectiveness and their intentions. Although the CDPs had failed to make any measurable impact on the living conditions of the inhabitants of the inner city, they had succeeded in making an indelible mark on the debate over their future. Small area action was widely perceived in local government and outside as wholly ineffective as a response to urban deprivation. When the Home Office in their last fling in their struggle for supremacy with the DoE launched their Comprehensive Community Programmes (CCPs), it was on the basis that no scale of action smaller than that of a complete local authority had any real chance of making an impact.

The special agencies created locally and nationally to deal with issues that fell outside the major welfare state functions also came under increasing criticism. The Home Office had in a White Paper on racial discrimination (Home Office, 1975) made one further attempt to move race relations issues from the periphery to the centre of attention of other spending departments. But the outcome fell short of their intentions and was confined to extending the scope of legislation against discrimination and combining the two bodies previously concerned with community relations and the implementation of the law into one new agency, the Commission for Racial Equality. Section 71 of the Race Relations Act 1976 that produced these outcomes did however also require local authorities to deal with discrimination within their area, a provision little noticed at the time in town halls, partly as a result of the total lack of enthusiasm displayed by the Department of the Environment. Nor did the creation of a parallel agency to deal with sexual discrimination, the Equal Opportunities Commission, impinge on the consciousness of local government at this stage.

All these developments helped to discredit the concept of co-ordinated and rational social planning. The failure of the joint approach to social policy and the attempt to produce priorities at the centre was matched by the inability of departmental planning systems to deliver sensible (that is usable) outcomes. As Glennerster comments:

> The learning element as far as central government was concerned was minimal. That should not obscure the fact that some health

authorities, local departments and corporate planning groups were beginning to make attempts to seek out client needs and measure them in an imaginative way, but the experiences were varied and the methods in an early stage of development even by 1980. For many critics the dominance of the centre and the weakness of the learning mode is no accident. (Glennerster, 1981, p. 43)

But none of these grounds for disillusionment was as convincing as the clear evidence of the ineffectiveness of past interventions. If urban decay in major cities had begun to emerge as an issue at the beginning of the decade, five years later the evidence was impossible to overlook. Even London, through the GLC, was now making insistent attempts to engage central government with the problems of acute social deprivation in Inner London. The rise of unemployment to levels of over 1 million, previously unknown in the postwar period, threatened to knock one of the props – the crucial one – away from under Beveridge's structure of welfare. However, that system was by now under more direct threat still. The landslip in British politics that took place after the election of Margaret Thatcher as Leader of the Conservative Party took the Conservatives away from the centre to the right and dissolved the postwar consensus on welfare, its organization and objectives. As late as 1973 Eversley could still write 'the backlash against the welfare state takes the relatively innocent form of an occasional Monday Club debate, a pathetic small crowd gathering behind some of our more extreme reactionary demagogues and the vocal but powerless men in the pubs and the clubs' (Eversley, 1973, p. 132). Five years later that analysis looked very thin: fundamental criticism of the postwar welfare state, from both left and right, had become commonplace, almost a new consensus.

This change had a good deal to do with the steady emergence of an effective right-wing critique from the Institute for Economic Affairs (IEA). The institute's approach to welfare was directed primarily to microeconomic objectives and centred mainly on the importance of securing and expanding consumer sovereignty. Fiscal measures that could ensure the better distribution of resources – of which Reverse Income Tax (RIT) was a prime example – could in their analysis push direct provision of welfare by state agencies back to the margin, where its extent could be further reduced by devices like abolition of rent control and the introduction of educational vouchers. Local government and its welfare services in this analysis figured only as agencies of central government – extreme examples of the general profligacy and inefficiency of the state.

In *Not From Benevolence*, their (pardonably) self-congratulatory celebration of the twentieth anniversay of the institute, Ralph Harris and Arthur Seldon set out the programme for the future of welfare in suitably hard-edged form. Radical reform would include:

> a) Charging for nursery, primary, adult education, G.P. and hospital services, rising to cover full costs over five years and adjusting step by step the support for low incomes and RIT;
> b) Raising present token charges for higher education to cover full tuition and maintenance costs which students could finance through loans repayable from resulting higher incomes;
> c) Charging for all central and local government services which do not meet the economic criterion of 'public goods'. (R. Harris and Seldon, 1977, p. 122)

Of the content of social policy, the definition of priorities as between groups in need, the quality of services delivered, there was little: where the consumer is sovereign whatever's chosen is best. Only in education did there appear to be some recognition that the debate did not finish with the method of payment.

If in 1977 much of this still seemed to those engaged in the field to be a cloud no bigger than a woman's hand, it was shortly to engulf the entire apparatus of the welfare state and send directors of social services scurrying desperately to the new tablets of stone (graven in Vienna) in search of some explanation of what was happening to them. But before these events occurred, there was time for one more major initiative in the field of poverty. In 1977 the new Secretary of State for the Environment, Peter Shore, produced a White Paper entitled *Policy for the Inner Cities* (Department of the Environment, 1977b). This document was based on the proposition that the key problem of the inner city was now economic decline: it also accepted that in attempting to cope with this issue any new initiative must be based on collaboration with local government. The device that it offered for securing such collaboration was that of partnership, and a three-year rolling programme jointly arrived at by the partners: central and local government and other statutory agencies. These programmes would be supplemented by resources from main programmes of all the participating agencies. Since the promotion of local economic initiatives and aid to industry required extension of local government powers, legislation – the Inner Urban Areas Act – was promoted. For those outside the magic circle of the seven partnerships a lower level of collaboration was provided for through 'programme' status. The Urban Programme, now finally removed from the Home Office, continued in new guise as the 'traditional' Urban Programme and a source of funding for individual projects (Higgins, 1983).

After some initial hesitation, local authorities agreed to participate in this new initiative, persuaded by the prospect of new resources on a substantial scale, the possibility of additional funding through the bending of main programmes of government agencies, and a promised diversion of the rate support grant towards urban authorities. The proposals had the additional merit, from the local government perspective, of building on initiatives already taken by individual authorities: Tyne and Wear, who had promoted their own legislation to engage in economic development; Rochdale, who had pioneered the concept of the industrial improvement area; and the GLC, which under the Labour administration of 1973–7 had begun a systematic programme for support for the London economy, based on the encouragement of co-operatives and small businesses and collaboration with other agencies – notably the Manpower Services Commission. Collaboration in partnership also offered the prospect of securing some modification in the policies of departments that impinged on local government – especially the regional policies of the Department of Industry and the Resource Allocation Working Party policies of the DHSS. But in practice neither central government nor local authorities proved able to deliver either on the modification of existing policies or the release of main programme resources: once again, the existing structures proved too strong.

One other feature of the 1977 initiative was the preparation of an appraisal of the needs of the area by the partnership authorities which formed part of the bid for resources. This approach reflected a general trend towards a more modest and specific setting of objectives in joint planning exercises, reflected also in the Housing Investment Programmes that the DoE had instituted. Some local authorities had gone further on their own account and attempted to produce coherent social strategy documents based around the broad theme of poverty, and providing both for direct interventions by the local authority concerned and the indirect involvement of other agencies like the Manpower Services Commission (MSC), the Area Health Authorities, or even in ambitious cases, the Supplementary Benefits Commission. One upper-tier authority that produced such a plan was Strathclyde Region; another was the GLC, who had acquired experience in this field running their own small-scale poverty initiative during the mid-1970s. The Strathclyde programme of 1976 had five basic elements: to bring pressure on local government and its agencies to deal with poverty; to tackle the severe problem of unemployment through the region's own efforts and by exerting pressure on central government; to encourage housing authorities to avoid concentrating disadvantaged families in areas of greatest need; to ensure that regional council and other services adequately provided for the needs of deprived areas; and to encourage the formation of robust and confident local community organizations (Hume, 1984).

Some of this activity was overtaken when the Conservatives won the 1979 election and began the processes of translating the concepts developed in opposition into executive action. Their approach to government represented a radical break with the past – and the whole range of local government activities in welfare were placed in a new context: for practitioners and advocates alike, the onus became to defend what was being attempted rather than to promote new activity.

A New Despotism?

Since government expenditure was identified as the crux of Britain's economic problems it is not surprising that one of the principal criteria on which the Conservative administration has assessed local government activity on social welfare has been its consequences for public expenditure. Since 1979 Treasury ministers have devoted considerable energy to attempts to impose restrictions on local government spending: their rationale for doing so was spelt out clearly by Leon Brittain in his speech made to local authority Chief Executives in July 1982. After rehearsing arguments about the legitimacy of central government concern with the quality of services and responsibility for 'consistency in standards', Mr Brittain concluded that:

> We are concerned to ensure that our policies designed to improve the health of the national economy are not frustrated by the actions of local authorities. This necessarily means that we are concerned to influence their decision about spending and the level of rates . . . local authorities are part of government as a whole. And their actions, like those of the rest of the public sector, can at their worst stifle enterprise, choke off growth and destroy jobs. (Jones and Stewart, 1984, p. 57)

The Treasury's general responsibility is for control of government expenditure as a whole. The major difference between the present government's approach and that of its predecessors is the closeness of concern and appetite for reductions. The outcomes for different local authority services are a function of the attitude of the spending departments responsible and their ministers. Thus housing, unprotected at ministerial level during Margaret Thatcher's first term, suffered the most severe cuts of all the social programmes, while expenditure on health remained protected by specific ministerial pledge. The impact elsewhere was very uneven: social services suffered some reductions, and the system of long-term planning was dismantled. Education appeared to present the opportunity of taking advantage of demographic trends to make reductions in both schools and higher education. The special programmes for the relief of poverty, the Urban Programme and the partnership arrangements established under the 1977 White Paper, were the subject of a special Treasury review but eventually survived almost unscathed in terms of level of expenditure. But modifications were made by government in the content and in the organization of the programme – notably the decision to hive off two partnership areas, the London Docklands and Liverpool, to Urban Development Corporations.

But changes in crude levels of expenditure conceal a variety of specific developments. The most pressing has been the question of the exercise of local discretion in choice of expenditure priorities, beginning with the legislation of 1980, the Local Government, Planning and Land Act. Here a variety of different imperatives have tugged in different directions. As with the 1951 Conservative government, considerable play was made with the lifting of a number of requirements for central supervision. But alongside this culling of obligations government has continued to lay new duties on local authorities – as, for example, over the sale of council housing – and to engage in detailed scrutiny of the content of policy, as in the case of the educational curriculum. At the same time successive secretaries of state for the Environment have been concerned to challenge the legitimacy of new activity undertaken by local government from the Urban Programme or through their own resources under section 137 of the Local Government Act 1972.

John Dearlove has argued that one direct consequence of the reform of local government after 1972 was to open it to local working-class demands (Dearlove, 1979, p. 257); it was some time before these were articulated in a distinctive form. But the conflicts between central and local government after the election of the Conservative government in 1979 has hastened the emergence of a new political class – younger Labour members conveniently but perhaps misleadingly personified by Ken Livingstone, elected Leader of the GLC by the Labour group after the 1981 GLC elections. Many of these Labour members had cut their teeth in local community organizations or in left-wing groups in opposition to the 'local state'. As Gyford shows (Chapter 4), they frequently serve in local government as officers, or local government trade unionists; alternatively, they may be full-time local politicians living on allowances and, where necessary, supplementary benefit. They conceive of their role as members in explicitly political terms and their accountability as belonging to the Labour movement as incarnated in the local party (often perceived as synonymous with the community) or with their trade union.

Their concern has been not principally with the social services in the traditional sense, except in those cases where they work in a particular service, where their main efforts have been directed to resisting cuts in jobs. Rather it has been with providing new forms of action: initiatives taken under section 71 of the 1976 Act – the creation of race relations committees and appointment of advisers; similar developments in the field of women's rights; attempts to create a system by which police forces are made more directly accountable to local government. Poverty strategies form part of this range of new activities: advocacy roles with other agencies and, in Strathclyde's case, action to maximize the income of local inhabitants by ensuring through advice that they receive their full entitlement to benefit and action to minimize the costs faced by low income households (Hume, 1984, p. 13). In all these initiatives there is a concern to secure the involvement of voluntary bodies whose aims are consistent with those of the authorities concerned: some authorities, notably the GLC, have built up a very substantial portfolio of support for such voluntary bodies. Finally, the new style of municipal socialism is committed, at least in principle, to decentralization of local government activity in the interests of popular accountability.

Despite the marked shift in the social policies of such authorities, there are a number of voluntary agencies that would view closer collaboration with considerable reserve. Hatch and Hadley in their polemic, *Social Welfare and the Failure of the State* (1981), state the case for disengagement from the traditional statutory sector; the case is further developed and refined by Francis Gladstone (1979). Such an approach to local social policy, homemade from the ground up, questions the validity of professionalism – even (or perhaps specially) in its community-oriented form. Instead it postulates 'authentic' self-help by communities with direct control over any projects that are sited in their area. The question of resources can best be resolved by seeking funds from some neutral untainted source; but many organizations are prepared to take funds direct from the vastly expanded range of central government initiatives – both DHSS and MSC have greatly increased the range and level of support for voluntary sector activity. Informal care in the community has been a particularly strong emphasis: self-help a general aspiration.

Some Conservative-controlled local authorities have been sympathetic to the new emphasis on the voluntary sector and support for volunteers in the DHSS over the past three years and have collaborated with initiatives designed to bring them into closer relationship with local social service departments. However, others have exhibited acute scepticism about any extension of publically funded activity in this area, and have seen their function as being to eliminate as much as possible, or where this is not feasible, to hive it off altogether. Sir John Gudgeon, leader of Kent County Council, told an IEA seminar that:

> We as members should never be afraid to face up to the possibility that the public could be better served if particular functions were to be undertaken in the private sector. We must examine whether there are any areas of service provision from which the local authority could withdraw, leaving market forces to fill the void, if any. (Gudgeon, 1980, p. 96)

How far any of these alternative approaches – the new municipal socialism, aggressive voluntarism, municipal monetarism or the attempt to create what has been termed 'a mixed economy of welfare' – will prove to be effective as responses to the social problems of deprived urban neighbourhoods now much exacerbated by five years of high unemployment, remains to be seen. When in 1981 poverty reappeared on the policy agenda – which, as Michael Heseltine observed in his Cabinet paper, 'it took a riot' to achieve – the measures that were taken in response were a modest increase in the allocation of resources to partnership, with special emphasis on economic projects and a rapid expansion in the activities of the MSC over a range of programmes directed at the young unemployed – youth training and work experience. Taken together with other new measures in the field of law and order – some of them stemming from Lord Scarman's Report (1981) on the Brixton disturbances and including some cautious experiments in the direction of local control – the Conservative government's urban policy might be categorized as containment, pending the long-deferred economic recovery.

Meanwhile local authorities are left with the rapidly expanding range of tasks confronting them in the social and economic arena in the mid-1980s: the consequences, direct and indirect, of persistent high unemployment (once again, to revive Slater's image, the rock on which society may split); local demographic changes including the increases in the number of the elderly and shifts in social patterns, notably the increase in the number of one-parent families. Alongside these major problems authorities have to cope with the additional responsibilities laid on them by a continuing flow of legislation – a recent and notorious example being responsibility for housing benefit and, especially in the case of inner city authorities, with government-imposed restrictions on expenditure.

Perhaps it is inevitable that what most local authorities, not all of them Labour controlled, see as a new despotism should at some stage encounter a new Poplarism, a refusal of consent – if only because of the practical impossibility of continuing to provide services. Meanwhile, the perpetually rising cost of providing social security has provoked the Secretary of State for the Social Services into launching a range of inquiries billed as the new Beveridge, and due to report by the time this volume is published. The secretary of state's critics have challenged him to be radical; but the likelihood is that his reviews will degenerate into another exercise in cost cutting and Treasury control. For local government, the new Bashaw is the Chancellor of the Exchequer.

11

Comparative Reflections and Projections

M. DAVID GELFAND

> A centenary should, however, be something more than an occasion for celebrating the triumphs of the past. A centennial anniversary affords also an opportunity for surveying the present situation and its problems, for inquiring into the process of contemporary development, for estimating whether the emerging trends seem worthy to succeed the achievements of the past. It may not, therefore, be inappropriate to conclude this volume with a chapter reviewing the major questions which loom large in the field of local government to-day and to indicate, however tentatively and briefly, the types of adjustment to which resort may be made in the effort to find satisfactory remedies for contemporary shortcomings.
>
> W. A. Robson in *A Century of Municipal Progress*, p. 455

A surprising characteristic of *Municipal Progress* is that a group of cosmopolitan authors wrote such an inward-looking book. Indeed, the few comparative assertions they do make often seem misguided or incomplete. It seems worthwhile, therefore, to conclude this successor volume with a chapter that attempts more direct comparative analysis of some of the issues raised in the preceding chapters.

Initially, however, I must record my own growing scepticism about unsystematic comparisons and transplantations of political institutions and processes. Casual comparisons can be misleading and self-serving at best and dangerous at worst. Moreover, the attempts made after 1935 to export some of the British local government patterns extolled in *Municipal Progress* – to West Germany, East Africa and the United States – had, at best, mixed results. Particularly complex adaptation problems arose when local government structures were imposed in colonial or postwar occupation settings without account being taken of the indigenous culture (Hiscocks, 1957; Mawhood, 1983a). As Mawhood observes: 'There is no such thing as a deductive theory of political decentralisation, working downwards from first principles to a logical prescription of the shape that local government authorities must take' (Mawhood, 1983b, p. 9). Kenya's experience, for example, illustrates the difficulties created by transplanting specific centre-periphery structures rather than general principles of local government administration (Oyugi, 1983). Nigeria's history chronicles the problems created by eclecticism – a semi-informed selection of bits and pieces of American, British and indigenous legal institutions without a coherent model for synthesizing the whole (Gboyega, 1983; Kasfir, 1983).

Anglo-American institutional exchanges have a long, but not always fruitful, history. One example of self-serving comparative analysis involved the approach of New York City reformers, in the early 1970s, to Greater London's local government system. After limited site visits, those who favoured political decentralization for New York found support for their preconceived position in the operation of the London boroughs (Task Force on Jurisdiction and Structure, 1972). Similarly, those who opposed decentralization found support for their preconceived position; they argued that Greater London's 1963–5 reorganization really involved metropolitanization coupled with substantial central government controls (City of New York Commission on City–State Relations, 1972). More recently, the enterprise zone concept has been wrenched from its Asian roots, poorly germinated in Britain, and now proposed for the United States with little or no analysis of its origins or of the need for adaptation (Callies and Tamashiro, 1983; Freilich, Frye and Carpenter, 1983). Nor have US institutions been easily transplanted in Britain.

On the other hand, carefully conducted and presented comparative analysis can assist both policy-makers and academic researchers in several ways. It can, for example, expand their universe of possibilities, both by raising fresh questions about their own governmental processes and by exposing them to a wider range of institutional alternatives drawn from other societies (Gelfand, 1975; Young and Kramer, 1978). It can also provide empirical data for testing some of the assumptions behind proposals for political, legal and economic reform and innovation. In this chapter the goals are somewhat narrower: essentially the aim is to explore, on a comparative basis, certain key themes that emerge from the preceding chapters and to identify certain political and legal factors which are unique to modern British local government and some factors which are shared with other Western societies.

This chapter will not attempt to subject all of the issues and cross-currents developed in the earlier chapters to comparative analysis; instead, it will discuss three key themes arising in those chapters and will locate them in a comparative context. The first theme is centralization versus decentralization, which raises issues of the legal status of local authorities, intergovernmental aid, central government administrative controls and local government resistance. The second theme is legal culture, especially the role of the courts in the regulation of local government affairs. The final theme is urban decline, with its concomitant mismatch between local needs and local resources.

Centralization Versus Decentralization

Though the earlier chapters in this volume analyse diverse aspects of British local government, nearly every one (like its counterpart in *Municipal Progress*) describes an ongoing struggle over the appropriate role of central government and of local government with respect to that chapter's particular functional or problem area. This issue of centralization versus decentralization has interesting comparative overtones.

In *Municipal Progress* Ivor Jennings argues that British local authorities 'are rigidly restricted to the powers conferred on them by Parliament; their organisation and their proceedings are determined by statute (there is no such thing as "home rule" in the American sense), and above all they are controlled more or less closely in all their activities by organs of the Central Government' (p. 450). The subordination of British local authorities to central government legal and administrative controls is also stressed by Robson (1966), and Jennings's perception that US local governments have greater autonomy is shared by some American authors (Foley, 1972; Sayre, 1973). Close analysis of the four critical factors – legal status, intergovernmental aid, controls over taxation and spending, and administrative regulation – reveals, however, that British local government was actually somewhat more autonomous than US local government until the dramatic changes introduced by the Local Government, Planning and Land Act 1980 (Wolman, 1982; Gelfand, 1983).

The legal status of local government should be analysed on a comparative basis, but simplistic terms and phrases provide precious little guidance. Thus, the constant repetition by the present government's ministers that England is 'a unitary and not a federal state' (Department of the Environment, 1983, p. 2; House of Lords, 1983; Jenkin, 1983, pp. 8–9, 22) has added very little to the debate over rate-capping. In fact, a unitary state may include local governments that are quite autonomous, for example, some Scandinavian countries (Hansen, 1981; Sharpe, 1981b), and a federal nation can have both autonomous local governments and fairly dependent ones. For example, local government structures and financial arrangements vary widely from state to state in the USA (Gelfand, 1979; Mandelker, Netsch and Salsich, 1983) and vary somewhat from one West German *Land* to another (N. Johnson and Cochrane, 1981). In order to evaluate the legal autonomy of a particular city within a federal system, however, one must begin by analysing the structure and operation of *both* federal *and* state controls. Consequently comparisons between British central–local relations and West German or American federal–local relations are necessarily incomplete.

Jennings's suggestion that 'home rule' guarantees US local government's broad autonomy (*Municipal Progress*, p. 450) is also incomplete. Several states have adopted statutory or constitutional home rule provisions, which have a dual purpose: to expand local government powers and to protect local governments from state government interference (Vandlandingham, 1968; Gelfand, 1983, p. 50). These provisions have reduced the number of judicial decisions employing Dillon's Rule (comparable to the British *ultra vires* principles) to invalidate local government action taken without specific statutory authorization. On the other hand, US local governments are constrained by federal constitutional guarantees of individual rights that have no counterparts in British central–local relations, for example, the Equal Protection Clause and Free Speech Clause (Robson, 1935a, p. 301; Gelfand, 1984). State constitutions contain similar provisions, sometimes interpreted to impose even greater constraints than the corresponding federal guarantees

(*Serrano* v. *Priest*). Other state constitutional provisions assign more specific duties to local governments such as backing their debt by full 'faith and credit' (*Flushing National Bank* v. *Municipal Assistance Corp.*).

Moreover, home rule has rarely served as an effective defence against state interference with local autonomy (Gelfand, 1981). Also, the 'state sovereignty' barrier erected by the US Supreme Court in *National League of Cities* v. *Usery*, to protect state and local governments from congressional interference, has been severely eroded (Gelfand, 1980, pp. 812–25, 839–49; 1983, p. 50). In short, no simple term or formula can adequately capture the similarities and differences in the legal status of local governments in different countries.

The effect of intergovernmental aid upon British local government autonomy has been the subject of perennial debate (Rhodes, 1981). Comparative analysis of this issue requires examinations of the level and growth of intergovernmental aid, variations from local authority to local authority within a country, and the nature (specific or general) of the intergovernmental aid. These detailed comparisons reveal that British intergovernmental aid, until the 1980 Act, has been somewhat less constraining than intergovernmental aid in the United States and other countries (Gelfand, 1981, 1983; Wolman, 1982; see also Hepworth, 1980; Sharpe, 1981a; Nathan, Doolittle and associates, 1983).

Control over its revenue sources and spending patterns is a critical aspect of a local government's autonomy. Britain and the Republic of Ireland are the only countries in Western Europe which still restrict their local authorities to a single tax source – the rates. Though many US and Canadian local governments also depend heavily upon taxes on real property, larger US cities and several Western European local authorities have added more cyclical, income-elastic taxes as revenue sources (Gelfand, 1981, pp. 671–81). Sweden and Norway have had positive experiences with the local income tax (Sharpe, 1981a). Though official and academic sources have repeatedly pointed to the many deficiencies of the rates (Layfield Committee, 1976, ch. 10; Foster, Jackman and Perlman, 1980; Department of the Environment, 1981; Prest, 1983), successive governments have refused to expand the revenue sources available to British local authorities, to reform the rating system or even to conduct needed property revaluations. Instead the present government pushed the Rates Act 1984 through Parliament over considerable opposition from members of all major parties, including many Tory back-benchers. This legislation, for the first time, authorizes the central government to limit the level of rates charged by English and Welsh local authorities.

US local governments (including home rule municipalities) have long operated under state-imposed limits upon their taxes on real property, with very restrictive limitations added in the late 1970s. These tax limits, however, are generally quoted as a maximum percentage of the assessed capital value of real property within the local government's boundaries. They apply to all localities in the state or, at least, to all localities within a population class (Gelfand, 1979). As a result, the calculation of the real estate tax limit for a particular local government does not require as much detailed involvement by central administrators as will likely occur under British rate-capping. Limitations on other forms of taxation may, however, be set by the state legislature or be increased through referenda on a locality-by-locality basis (Gelfand, 1981; Wolman, 1982).

A few US states impose constitutional or statutory limits upon expenditure by local governments, and several state constitutions limit growth in state government budgets. These limitations, however, exclude some forms of spending, allow budgets to grow by a set percentage each year, or permit increases in line with growth in population or personal income (Gelfand, 1979). Attempts to control English local authority spending indirectly, by reducing the rate support grant to 'overspenders', has had only limited success (Jenkin, 1983, pp. 12–13; Rhodes, 1984). However, the British central government can now wield enormous power over local spending, by combining its control over the principal form of intergovernmental aid with its new control over the sole form of local taxation. There is also more central government involvement in calculating and allocating each local authority's capital expenditure limit, under the 1980 Act, than is necessary for setting the debt limits of most US municipalities (Gelfand, 1981).

Administrative controls by central government departments over British local authority activities evoked criticism by Jennings in *Municipal Progress* (p. 450). Whatever the status of administrative regulation in 1935, when Jennings was writing, the 1946–79 pattern of British central–local relations seemed to involve the exercise of controls only at key pressure points coupled with administrative persuasion and financial incentives (Griffith, 1966; Greenwood,

1981; Jones and Stewart, 1982). This pattern contrasts sharply with the direct, daily administrative and legal controls employed by the French prefectoral corps and by many US state government agencies during the same period. Interestingly, France has recently moved rapidly away from this legal-administrative approach and towards decentralization (Mény, 1984), while Britain has increased central government's grip over local authorities. In fact, several commentators have argued that the 'fundamental' rules governing Britain's central–local relationship have been rewritten by provisions of recent legislation – the Local Government, Planning and Land Act 1980, Local Government (Miscellaneous Provisions) (Scotland) Act 1981, Local Government Finance Act 1982, and Rates Act 1984 (Elliott, 1981; G. W. Jones and Stewart, 1983; Loughlin, 1983). West German leaders at all levels appear to have sought a more 'co-operative' intergovernmental system, maintained through a fair amount of legal, financial and administrative decentralization (N. Johnson and Cochrane, 1981; von Beyme, 1983).

The current emphasis upon conflict in intergovernmental relations should not blind observers to the fact that local governments in Western countries regularly co-operate with central programmes and generally comply voluntarily with central government requests (Hanf, 1978). Co-operation and compliance may result from cooption in particular cases, general party loyalty, or a shared 'operating ideology' or 'assumptive world' (Young, 1977; Sharpe, 1983).

Local government resistance to central government pressures can, however, take several different forms. Local government politicians and officials in most Western countries can often blunt, deflect or redirect unacceptable central policies at the formative stage. Resistance can be especially effective if local authorities have direct or indirect representation in the national legislature. Direct representation, like that of the West German *Länder* in the Bundesrat, is rare in societies committed to popular elections. Indirect representation is possible where legislative election districts coincide with the boundaries of lower-level governments. For example, all US senators and the congressmen from a few small states are elected from districts that coincide with state boundaries. Such overlapping of election districts is rare at the state level because the one-person-one-vote Equal Protection standard (*Reynolds* v. *Sims; Avery* v. *Midland County*) permits only limited deviations from population equality for state legislative election districts which follow local government boundaries (*Mahan* v. *Howell; Brown* v. *Thomson*; Gelfand, 1984).

Alternatively, national politicians may be attuned to local political concerns as a result of their present or past involvement in local politics. For example, under France's *cumul des mandats* tradition, many local elected leaders simultaneously hold national office (Ashford, 1982; Mény, 1984). By contrast, multiple office holding of this sort is exceedingly rare in Britain (Ashford, 1982; Sharpe, 1983, p. 5), and it is unlawful in many US states (*Arceneaux* v. *Treen*). However, countless current national and state politicians and officials have previously held elected or appointed positions at lower levels of US and West German government. Such mobility among levels of government is somewhat less common for British politicians, and it is exceedingly rare for British civil servants.

Local governments can attempt to influence central government legislative or administrative policy by lobbying through their national associations (Rhodes, Hardy and Pudney, 1983a), political parties (Gyford and James, 1983; Gyford, Chapter 4) or 'policy communities' (Wright, 1974; Rhodes, Hardy and Pudney, 1983b). US local government politicians and officials regularly lobby state legislators and members of Congress. This sort of direct lobbying by local authorities is gradually emerging in Britain. It failed to gain concessions for local government in the Local Government, Planning and Land Act 1980. It was somewhat more productive with respect to the Local Government Finance Act 1982, and it was crucial to the defeat (in the House of Lords) of key provisions of the Local Government (Interim Provisions) Bill 1984.

If local governments fail to win sufficient concessions during the formation of a legislative or administrative policy, they may still employ various strategies to prevent or delay the implementation of that policy. Particular local leaders may knowingly engage in overt, illegal protest actions (Alford, 1969, pp. 287–8). Examples are the resistance by some US state and local politicians to federal court desegregation decrees (*Cooper* v. *Aaron*) and the Poplar (Branson, 1979) and Clay Cross (Tavern, 1976) protests in Britain. Local politicians may also seek to encourage public opposition to central government policies. Some US, English and Scottish local leaders have adopted this approach with respect to state or national fiscal policies. More commonly, local leaders can quietly delay or redirect a centrally mandated policy. They often possess a near monopoly of crucial information, and they usually have the primary responsibility for policy execution (Hanf, 1978; Rhodes, Hardy and Pudney, 1983b; Young, 1983). Finally, local governments may seek to enlist the assistance of the courts in their opposition to unpalatable central government policies. This is a new strategy for left-wing Labour councillors, who have long distrusted the bench and the bar and have regarded judicial activity as part of the centralizing process. US, French and West German local authorities have a longer history of employing the courts both as a sword and as a shield.

Lawyers, the Courts and the Law

Lawyers are more prevalent and generally play more prominent parts in the intergovernmental dramas in France, West Germany and the USA than do their counterparts in British central–local government relations. The existence of constitutional judicial review in the first three countries, and the more overt articulation of policy positions by courts which generally accompanies it, are clearly important explanatory factors.

Law is the most common form of professional training for high-level appointed positions at the federal and state levels of US and West German government. Lawyers are not quite so numerous at the local government level, especially in smaller towns. The situation in Britain is virtually the reverse. There are relatively few barristers or solicitors in the civil service and most of them play a limited, technical role (like that of architects or engineers). By contrast, legal training has long been considered essential for the generalists of British local government – first the Town Clerks and now City Secretaries and many Chief Executives (Headrick, 1962: Drewry, 1981; Elliott, 1981; Gelfand, 1981; J. D. Stewart, 1983).

Scholarly writing on British local government was long dominated by legal academics and others with a legal-administrative perspective, including several of the authors of *Municipal Progress* itself (Chester, 1951; Jennings and Griffith, 1960; Griffith, 1966; Robson, 1966). By the mid-1960s, however, British legal academics, like the practising bar, were concentrating upon particular local government activities, such as planning and housing (Arden, 1980; Grant, 1983). Very few legal treatises or articles in the 1970s considered the wide range of local authority services and activities or the overall legal structure of central–local government relations.

The current period of flux, doubt and confrontation in British central–local relations seems to call upon lawyers, courts and legal academics to play somewhat broader roles than they have in the recent past or are likely to in the future. British courts have regularly entertained cases involving local authorities (Cross, 1981, pp. 1–18; K. Davies, 1983, pp. 185–7; R. Jones, 1981). However, they have very limited experience with cases in which local authorities sue each other, such as *Bromley LBC* v. *Greater London Council*, or central departments and local authorities confront one another, such as *R.* v. *Secretary of State for the Environment, ex parte Brent LBC*. These new cases with greater political salience seem to be the product of several factors: dwindling public and private financial resources, as described by Dawson (Chapter 2); increased confrontational style among central and local government politicians, as described by Gyford and Deakin (Chapters 4 and 10); and changes in court personnel and procedures which facilitate public law litigation. Lawyers (Marks, 1982; Sofer, 1983) and legal academics (Elliott, 1981, 1982; Loughlin, 1982, 1983) are needed as intermediaries because the courts have an intellectual history, set of values and work patterns which are different from those of the traditional political arenas in which most local authority councillors operate. Also judges need to be educated about the difficult political and legal issues raised by this new breed of cases (cf. Griffith, 1977).

Intergovernmental suits occur somewhat more frequently in the USA, and they are not generally regarded as reflecting a breakdown in the intergovernmental system (*National League of Cities* v. *Usery; Massachusetts* v. *United States; State* v. *City of Sunrise; City of New Orleans* v. *Treen*). In part, this difference in view is the result of different cultural attitudes about litigation and the negotiation of disagreements. That is, British politicians, at least until recently, have believed that central–local disputes should be resolved behind closed doors rather than aired in public. Their US and West German counterparts tend to see intergovernmental relations as the product of a written corporate charter, whose terms can be renegotiated or interpreted by the courts as necessary.

Constitutional and administrative courts are quite important to political life in France, West Germany and the USA. On the whole, the civil rights litigation in the USA employs causes of action pursued by private individuals in the federal and state courts, with the full panoply of judicial remedies – declaratory relief, injunctions, damages and attorney's fees and costs (Gelfand, 1984). These suits are often based upon provisions of the United States Constitution, especially the First, Fourth and Fourteenth Amendments. However, many civil rights claims are premised upon federal statutes, especially the Civil Rights Act of 1964 (Eisenberg, 1981). In Britain remedies for discrimination must generally be pursued through mediation and conciliation, with some resort to administrative tribunals.

The Experience of Urban Decline

Much of the optimism in *Municipal Progress* springs from the authors' experience of growth and prosperity in British cities, but several of the chapters in this book reflect a different economic pattern of urban decline and fiscal pressure. Though the particular causes, details and political responses may vary from country to country, a comparable mismatch between urban needs and resources has occurred in many Western nations. This needs–resources mismatch is sometimes characterized as 'the urban fiscal crisis', a subject which dominated the transatlantic literature during the 1970s and early 1980s (O'Connor, 1973; US Advisory Commission on Intergovernmental Relations, 1973; Gelfand, 1981; Sharpe, 1981a; Wolman, 1981). These urban economic and fiscal problems have especially severe impacts upon minority groups, who are disproportionately located in older inner cities.

Large cities on both sides of the Atlantic are under severe pressure to provide more specialized, expensive services, as a result of demographic changes, declining private sector employment opportunities, public sector labour union demands and service mandates by higher levels of government. At the same time, they are faced with an eroding tax base, as a result of demographic and industrial changes and taxpayer resistance (Gelfand, 1979, pp. 555–8; 1981, pp. 690–1). Moreover, politicians at higher levels of government often fear and resent the actual and potential influence of the political leaders of such key cities as London, Paris and New York (or other US cities with a dominant role in their states). Consequently, the legal and financial authority of these cities has often been restricted

(Gelfand, 1981, pp. 664–7), and their highly publicized recent financial difficulties have sometimes led to still further constraints.

Dramatic changes in population and employment opportunities, especially in manufacturing industries, appear to be responsible for Britain's recent urban decline. Until the 1950s the pattern has been urban growth and concentration. But during that decade the pace of growth moderated in London, Merseyside, Manchester and the West Midlands. In the 1960s and 1970s these key metropolitan areas faced both population and employment declines, while non-metropolitan areas, especially portions of the south-west and East Anglia, recorded net gains (Hall and Hay, 1980; McCallum, 1980; Young and Mills, 1983).

'Counter-urbanization' has been even more extreme in the USA, with the mutually reinforcing migrations of population and employment from inner cities to suburban and fringe areas, coupled with a substantial regional shift from the older industrial cities of the north-east and north central states to the south and west (Berry and Fillard, 1977; Sternlieb and Hughes, 1977; Weinstein and Firestine, 1978; Hall and Hay, 1980). In both Britain and the USA,

> the forces for decentralisation so apparent during the 1970s have been linked with accelerated decline of the central areas of the largest cities. Migration has been highly selective, and the central city areas have come to be populated increasingly by those who do not have the option to leave. (Young and Mills, 1983, p. 38)

Similar patterns are emerging in West Germany, Switzerland, Austria and the Low Countries (Hall and Hay, 1980; Fielding, 1982), but foreign immigration and a relative lag in economic development appear to have delayed counter-urbanization in other parts of Western Europe, Canada and Australia (Hall and Hay, 1980, pp. 17, 227–8; Young and Mills, 1983).

The linked processes of metropolitan decline and non-metropolitan area growth mean that younger, middle-class and professional persons leave inner city areas while poorer, less-skilled and public-sector-dependent persons (elderly, very young, single-parent families, minorities) remain. These demographic trends have serious implications for the fiscal health of central city governments and for racial minorities remaining in the inner cities (Blaydon and Gilford, 1976; Temporary Commission on City Finances, 1978, pp. 22, 59;

Young and Mills, 1983, p. 38). Governmental responses to these demographic changes have included: local government reorganization schemes designed to expand the tax base for US inner city governments and to improve both service delivery patterns and partisan positions, in both the USA and Britain (Alexander, Chapter 3; Foley, 1972, pp. 177–9, 185–9; Leiken, 1972); targeted central government grants, which were later diluted, for example, the USA's Model Cities Program and Britain's Urban Programme (Deakin, Chapter 10); economic development programmes and tax incentives intended to encourage private investments.

Comparisons between race relations in any two societies, especially Britain and the USA, must take account of the vast differences in the proportions and history of minority groups in the two countries. Race and ethnicity have long been important political issues in most US states and cities and in some Canadian provinces and cities; they have attained national prominence there today. The political salience of race in these two societies reflects their multi-ethnic history. Though regional and 'national' (Scottish, Welsh and Irish) cultural conflicts have been part of British political life for some time, both local and national politics have traditionally operated on the assumption that Britain is – or should be – an ethnically homogenous society. As the number of immigrants and economic pressures have increased, however, race and ethnicity have emerged as important social issues and have become an important part of local politics in some British cities (Gyford, Chapter 4; Deakin, Chapter 10).

Despite these important differences in numbers and history, arising from post-slavery developments in the one case and post-colonial immigration in the other, British journalists and scholars have looked across the Atlantic for explanations of and responses to the recent wave of urban riots (Glazer and Young, 1983). Indeed, a similar pattern has developed on both sides of the Atlantic – official indifference, followed by temporary and heavily publicized political and 'blue ribbon panel' attention to 'the problem', followed by a temporary shift of financial resources, followed by institutionalization of limited legal-administrative responses (Benyon, 1983). Interestingly the responses to racial inequality have had centralizing overtones in both countries, with new specific grants and administrative regulations emerging in Britain (Deakin, Chapter 10) and civil rights legislation authorizing federal court intervention in local government affairs in the USA (Gelfand, 1984).

Conclusion

In conclusion then, there are significant parallels and contrasts between Britain and other societies. Comparative analysis throws new light upon the key issues of centralization and decentralization, the role of the courts and change in the urban system. But whereas the challenge posed by *urban* decline and the resulting fiscal pressures must be faced, it need not entail *municipal* decline. The US, French and West German experience demonstrates that greater centralization is not an inevitable result of the changing context of local government. It is, rather, a matter of clear political choice.

Bibliography

Abel-Smith, B. and Townsend, P. B. (1965), *The Poor and the Poorest: a New Analysis of the Ministry of Labour's Family Expenditure Surveys of 1953–4 and 1960*, Bell.

Adam Smith Institute (1983), *The Omega File: Local Government, Planning and Housing*, Adam Smith Institute.

Alexander, A. (1982a), *The Politics of Local Government in the United Kingdom*, Longman.

Alexander, A. (1982b), *Local Government in Britain since Reorganisation*, Allen & Unwin.

Alford, R. A. (1969), 'Explanatory variables in the comparative study of urban administration and politics', in R. T. Daland (ed.), *Comparative Urban Research: The Administration and Politics of Cities*, Sage.

Alt, J. (1971), 'Some social and political correlates of county borough expenditures', *British Journal of Political Science*, vol. 1, no. 1.

Ambrose, P. and Colenutt, B. (1975), *The Property Machine*, Penguin.

Arden, A. (ed.) (1980), *Encyclopaedia of Housing Law and Practice*, Sweet & Maxwell.

Ashford, D. E., (1982), *British Dogmatism and French Pragmatism*, Allen & Unwin.

Ashworth, W. (1960), *An Economic History of England, 1870–1939*, Methuen.

Bains, L. (Chairman), (1972). *The New Local Authorities: Management and Structures*, HMSO.

Ball, M. (1983) *Housing Policy and Economic Power*. Methuen

Barcley, P. M. (Chairman) (1982), *Social Workers: Their Role and Tasks*, Report of a Working Party for the National Institute for Social Work, Bedford Square Press.

Barlow, Sir M. (Chairman) (1940). *Report of the Royal Commission on the Distribution of the Industrial Population*, Cmd 6153, HMSO.

Barras, R. (1985) 'Redevelopment of profit and development control', in S. M. Barrett and P. Healey (eds), *Land Policy: Problems and Alternatives*, Gower.

Barrett, S. and Fudge, C. (eds) (1981), *Policy and Action*, Methuen.

Barrett, S. and Hill, M. (1984), 'Policy, bargaining and structure in implementation theory: towards an integrated perspective', *Policy and Politics*, vol. 12, no. 3.

Bassett, R. M. (1958), *1931: Political Crisis*, Macmillan.

Beales, H. L. (1936), 'A Century of Municipal Progress', *Political Quarterly*. January–March.

Beer, S. (1965), *Modern British Politics*, Faber.

Beloff, M. (1975), 'Introduction', in G. Peele and C. Cook (eds), *The Politics of Reappraisal 1918–1939*. Macmillan.

Benyon, J. (ed.) (1983), *Scarman and After*, Pergamon.

Berry, B. J. L. and Fillard, Q. (1977), *The Changing Shape of Metropolitan America: Commuting Patterns, Urban Fields, and Decentralization Processes*, Ballinger.

Best, R. H. (1977), 'Agricultural land loss – myth or reality', *The Planner*, vol. 63, Jan.

Beveridge, Sir W. (1942), *Report on Social Insurance and Allied Services*, Cmnd 6404, HMSO.

Blaydon, C. C. and Gilford, S. R. (1976), 'Financing the cities: an issue agenda', *Duke Law Journal*, p. 1057.
Blom-Cooper, L. (1982), 'The new face of judicial review: administrative changes in Order 53', *Public Law*, Summer.
Blunkett, D. and Green, G. (1983), *Building from the Bottom*, Fabian Society.
Boaden, N. (1971), *Urban Policy-Making*, Cambridge University Press.
Boddy, M. and Fudge, C. (eds) (1984), *Local Socialism?* Macmillan.
Bowley, M. (1945), *Housing and the State 1919–1944*, Allen & Unwin.
Branson, N. (1979) *Poplarism 1919–1925*, Lawrence & Wishart.
Briggs, A. (1968), *Victorian Cities*, Penguin.
Bristow, S. (1978), 'Local politics after reorganisation – the homogenisation of local government in England and Wales', *Public Administration Bulletin*, no. 28.
Bulpitt, J. (1967), *Party Politics in English Local Government*, Longman.
Bulpitt, J. (1983), *The Territorial Policies of the Thatcher Government 1979–1983*, PSA Workgroup on United Kingdom Politics, Oxford.
Burgess, T. and Travers, T. (1980), *Ten Billion Pounds*, Grant McIntyre.
Butler, D. and Stokes, D. (1975), *Political Change in Britain*, Macmillan.
Callies, D. L. and Tamashiro, G. M. (1983), 'Enterprise zones: the redevelopment sweepstakes begins', *Urban Lawyer*, vol. 15, no. 1.
Cannadine, D. (ed.) (1982), *Patricians, Power and Politics in Nineteenth Century Towns*, Leicester University Press.
Cannan, Edwin (1912), *The History of Local Rates in England*, 2nd edn, King.
Carlson, R. E. (1947), *British Block Grants and Central–Local Finance*, Johns Hopkins University studies in Historical and Political Science, series 65, no. 1, Johns Hopkins Press.
Central Policy Review Staff (1975), *A Joint Framework for Social Policy*. HMSO.
Chayes, A. (1976), 'The role of the judge in public law litigation', *Harvard Law Review*, vol. 89, no. 7.
Checkland, S. (1983), *British Public Policy 1776–1939*, Cambridge University Press.
Cherry, Gordon (1982), *The Politics of Town Planning*, Longman.
Chester, D. N. (1951), *Central and Local Government*, Macmillan.
City of New York Commisson on City–State Relations (1972), *Decentralizing New York City Government: The London Experience – Misunderstood?*
Cockburn, C. (1977), *The Local State*, Pluto.
Cole, G. D. H. (1948), *A History of the Labour Party from 1914*, Routledge & Kegan Paul.
Community Development Project (1977), *Gilding the Ghetto*, Home Office.
Cook, C. (1975a), *The Age of Alignment*, Macmillan.
Cook, C. (1975b), 'Liberals, Labour and local elections', in G. Peele and C. Cook (eds), *The Politics of Reappraisal 1918–1939*, Macmillan.
Cousins, P. (1982), 'Quasi-official bodies in local government', in A. Barker (ed.), *Quangos in Britain*, Macmillan.
Craig, P. P. (1983), *Administrative Law*, Sweet & Maxwell.
Crawford, C. and Moore V. (1983), *The Free Two Pence*, Chartered Institute of Public Finance and Accountancy.
Crosland, C. A. R. (1956), *The Future of Socialism*, Cape.
Cross, C. A. (1981), *Principles of Local Government Law*, 6th edn, Sweet & Maxwell.
Crossman, R. H. S. (1975), *The Diaries of a Cabinet Minister, Vol. I: Minister of Housing and Local Government*, Hamish Hamilton/Cape.
Cullingworth, J. B. (1966), *Housing and Local Government*, Allen & Unwin.

Cullingworth, J. B. (1975), *Environmental Planning 1939–1969: Vol. 1, Reconstruction and Land Use Planning 1939–1947*, HMSO.
Davies, J. G. (1972), *The Evangelistic Bureaucrat*, Tavistock.
Davies, K. (1983), *Local Government Law*, Butterworths.
Davies, M., Game, C., Gibson, J. and Stewart, J. D. (1983), *Grant Characteristics and the Budgeting Process*, Institute of Local Government Studies.
Deakin, N. (1982), 'Research and the policy-making process in local government', *Policy and Politics*, vol. 10, no. 3.
Dearlove, J. (1973), *The Politics of Policy in Local Government*, Cambridge University Press.
Dearlove, J. (1979), *The Reorganisation of British Local Government*, Cambridge University Press.
Denning, A. (1949), *Freedom Under the Law*, Stevens.
Dennis, N. (1970), *People and Planning*, Faber.
Department of Education and Science (1972), *Education: A Framework for Expansion*, Cmnd 5174, HMSO.
Department of the Environment (1971), *The Future Shape of Local Government Finance*, Cmnd 4741, HMSO.
Department of the Environment (1977a). *Housing Policy*, Cmnd 6851, HMSO.
Department of the Environment (1977b), *Policy for the Inner Cities*, Cmnd 6845, HMSO.
Department of the Environment (1981), *Alternatives to Domestic Rates*, Cmnd 8449, HMSO.
Department of the Environment (1983), *Rates*, Cmnd 9008, HMSO.
Devereux, M. and Morris, N. (1984), 'The Chancellor's arithmetic', *Fiscal Studies*, vol. 5, no. 2.
Devlin, P. (1956), 'Public policy and the executive', *Current Legal Problems*, vol. 9.
Dicey, A. V. (1885), *Introduction to the Study of the Law of the Constitution*, Macmillan.
Donnison, D. (1983), *Urban Policies: a New Approach*, Fabian Society.
Donoughmore Report (1932), *Report of the Committee on Ministers' Powers*, Cmd 4060, HMSO
Donoughue, Bernard and Jones, G. W. (1973), *Herbert Morrison: Portrait of a Politician*, Weidenfeld & Nicolson.
Drewrey, G. (1981), 'Lawyers in the U.K. civil service', *Public Administration*, vol. 59, spring.
Dunham, A. (1964), 'Property, city planning and liberty', in M. Stewart (ed.) (1972), *The City: Problems of Planning*, Penguin.
Dunleavy, P. (1979), *Urban Political Analysis*, Macmillan.
Dunleavy, P. (1981), *The Politics of Mass Housing in Britain, 1945–1975*, Oxford University Press.
Dworkin, R. (1977), *Taking Rights Seriously*, Harvard University Press.
Eisenberg, T. (1981), *Civil Rights Legislation*, Bobbs-Merrill.
Elcock, H. (1981), 'Tradition and change in Labour Party politics: the decline and fall of the city boss', *Political Studies*, vol. 39, no. 3.
Elias, N. (1978), *The Civilising Process: The History of Manners*, Blackwell.
Elias, N. (1982), *The Civilising Process: State Formation and Civilisation*, Blackwell.
Elliott, J. (1971), 'The Harris Experiment in Newcastle upon Tyne', *Public Administration*, vol. 49, summer.
Elliott, M. (1980), 'ACAS and judicial review,' *Modern Law Review*, vol. 43, No. 5.
Elliott, M. (1981), *The Role of Law in Central–Local Relations*, SSRC

Elliott, M. (1982), 'Local government in Schleswig-Holstein – the intrusion of law', *Local Government Studies*, vol. 8, Jan./Feb.
Elliott, M. (1983), 'Constitutional continuity and the position of local government', in K. Young (ed.), *National Interests and Local Government*, Heinemann.
Ermisch, J. (1984), *Housing Finance: Who Gains?*, Policy Studies Institute.
Evans, A. (1976), 'Capital valuation for domestic rating', in Layfield (1976), Appendix 10.
Eversley, D. (1973), *The Planner in Society*, Faber.
Fielding, A. J. (1982), *Counter-urbanization in Western Europe*, Pergamon.
Finer, H. (1933), *English Local Government*, Methuen.
Finer, H. (1941), *Municipal Trading*, Allen & Unwin.
Fogarty, M. F. (1948), *Town and Country Planning*, Hutchinson.
Foley, D. L. (1972), *Governing the London Region: Reorganization and Planning in the 1960s*, University of California Press.
Foot, M. (1975), *Aneurin Bevan: Vol. II 1945–1960*, Paladin.
Forsyth, M. (1981), *Re-servicing Britain*, Adam Smith Institute.
Foster, C. D., Jackman, R. A. and Perlman, M. (1980), *Local Government Finance in a Unitary State*, Allen & Unwin.
Fothergill, S. and Gudgin, G. (1982), *Unequal Growth: Urban and Regional Employment Change in the U.K.*, Heinemann.
Fraenkel, E. (1941), *The Dual State*, Oxford University Press.
Franks (Chairman) (1957), *Report of the Committee on Administrative Tribunals and Enquiries*, Cmnd 218, HMSO.
Fraser, D. (1979), *Power and Authority in the Victorian City*, Blackwell.
Freilich, R. H., Frye, D. S. and Carpenter, D. T. (1983), 'The new federalism – American urban policy in the 1980's: trends and directions in urban, state, and local government law', *Urban Lawyer*, vol. 15, no. 1.
Friedman, M. (1962), *Capitalism and Freedom*, University of Chicago Press.
Fudge, C. (1981), 'Winning an election and gaining control: the formulation and implementation of a local political manifesto', in S. Barrett and C. Fudge (eds), *Policy and Action*, Methuen.
Fudge, C. (1984), 'Decentralisation: Socialism goes local?', in M. Boddy and C. Fudge (eds), *Local Socialism?* Macmillan.
Fulton, Lord (Chairman) (1968), *Report of the Committee on the Civil Service, 1966–68*, Vol. 1, Cmnd 3638, HMSO.
Game, C. and Skelcher, C. (1983), 'Manifestos and other manifestations of local party politics', *Local Government Studies*, vol. 9, no. 4.
Gboyega, A. (1983), 'Local government reform in Nigeria', in P. Mawhood (ed.), *Local Government in the Third World: The Experience of Tropical Africa*, Wiley.
Gelfand, M. D. (1975), 'Conflict resolution in a politically decentralized local government system', *Columbia Journal and Law and Social Problems*, vol. 2, no. 4.
Gelfand, M. D. (1979), 'Seeking local government financial integrity through debt ceilings, tax limitations, and expenditure limits: the New York City fiscal crisis, the taxpayers' revolt, and beyond', *Minnesota Law Review*, vol. 62, no. 4.
Gelfand, M. D. (1980), 'The Burger Court and the new federalism: preliminary reflections on the roles of local government actors in the political dramas of the 1980's', *Boston College Law Review*, vol. 21, no. 4.
Gelfand, M. D. (1981), 'Comparative urban finance: are the London and Brooklyn bridges falling down?' *Tulane Law Review*, vol. 55, no. 3.
Gelfand, M. D. (1983), 'Comment on Chapter 3', in K. Young (ed.), *National Interests and Local Government*, Heinemann.

Gelfand, M. D. (1984), *Federal Constitutional Law and American Local Government: A Treatise for City Attorneys, Public Interest Litigators, and Students*, Michie.

Gibbon, I. G. (1931), 'Some problems of local government', *Public Administration*, April.

Gibson, J. G. (1982), 'The block (and target) grant system and local authority expenditure – theory and evidence', *Local Government Studies*, vol. 8, no. 3, May/June.

Gilbert, B. (1980), *Britain Since 1918*, Batsford.

Gillard, M. and Tomkinson, M. (1980), *Nothing to Declare: the Political Corruption of John Poulson*, John Calder.

Gladstone, Francis (1979), *Voluntary Action in a Changing World*, Bedford Square Press.

Glazer, N. and Young, K. (eds) (1983), *Ethnic Pluralism and Public Policy*, Heinemann.

Glennerster, H. (1981), 'From containment to conflict? Social planning in the seventies', *Journal of Social Policy*, vol. 10, pt 1.

Glennerster, H. (ed.) (1983), *The Future of the Welfare State: Remaking Social Policy*, Heinemann.

Glennerster, H. with Korman, N. and Marslen-Wilson, F. (1983), *Planning for Priority Groups*, Martin Robertson.

Goddard, J. B. (1983), 'Structural change in the British space economy', in J. B. Goddard and A. G. Champion (eds), *The Urban and Regional Transformation of Britain*, Methuen.

Gosden, P. (1983), *The Education System Since 1944*, Martin Robertson.

Grant, M. (1983), *Urban Planning Law*, Sweet & Maxwell.

Greater London Council (1969), *Tomorrow's London: a Background to the Greater London Development Plan*, Greater London Council.

Greater London Council (1984), *Responses by GLC Committees to the Government's White Paper 'Streamlining the Cities'*, Greater London Council.

Green, D. G. (1981), *Power and Party in an English City*, Allen & Unwin.

Greenwood, R. (1981), 'Fiscal pressure and local government in England and Wales', in C. Hood and M. Wright (eds), *Big Government in Hard Times*, Martin Robertson.

Greenwood, R., Norton, A. L. and Stewart, J. D. (1969), *Recent Reforms in the Management Arrangements of County Boroughs in England and Wales*, Institute of Local Government Studies.

Greenwood, R. and Stewart, J. D. (1974), *Corporate Planning in Local Government: an Analysis with Readings 1969–1972*, C. Knight.

Griffith, J. A. G. (1959), 'The law of property (land)', in M. Ginsberg (ed.), *Law and Opinion in England in the Twentieth Century*, Stevens.

Griffith, J. A. G. (1966), *Central Departments and Local Authorities*, Allen & Unwin.

Griffith, J. A. G. (1977), *The Politics of the Judiciary*, Fontana.

Griffith, J. A. G. (1979), 'The political constitution', *Modern Law Review*, vol. 42, no. 1.

Griffith, J. A. G. (1981), *Public Rights and Private Interests*, Academy of Legal Publications, Trivandrum.

Gudgeon, Sir J. (1980), 'How far can local government act with real autonomy if financed largely by central government?', in D. N. King and D. Anderson (eds) *Town Hall Power or Whitehall Pawn*, Institute of Economic Affairs.

Gunby, D. S. (1984) 'Town planning and corporate power between the wars', in

S. V. Ward (ed.), *Planning and Economic Change: A Historical Perspective*, Oxford Working Paper 78, Town Planning Department, Oxford Polytechnic.
Gyford, J. (1984), *Local Politics in Britain*, 2nd edn, Croom Helm.
Gyford, J. and James, M. (1983), *National Parties and Local Politics*, Allen & Unwin.
Haar, C. M. (1951), *Land Planning in a Free Society*, Harvard University Press.
Hadow, W. H. (Chairman) (1934), *Qualifications for Recruitment, Training and Promotion of Local Government Officers*, Report of a Departmental Committee, HMSO.
Hall, P. (ed.) (1981), *The Inner City in Context*, Heinemann.
Hall, P. and Hay, D. (1980), *Growth Centres in the European Urban System*, University of California Press.
Hammond, J. L. (1935), 'The social background: 1835–1935', in H. J. Laski, W. I. Jennings and W. A. Robson (eds), *A Century of Municipal Progress 1835–1935*, Allen & Unwin.
Hampton, W. (1970), *Democracy and Community*, Oxford University Press.
Hampton, W. (1971), 'Little men in big societies', *Journal of the Royal Town Planning Institute*, vol. 57, no. 4.
Hampton, W. (1980), 'Introduction', in *Party Politics in Local Government: Officers and Members*, Policy Studies Institute and Royal Institute of Public Administration.
Hanf, K. (1978) 'Introduction', in K. Hanf and F. Schapf (eds), *Interorganizational Policymaking: Limits to Coordination and Central Control*, Sage.
Hanham, H. J. (1959), *Elections and Party Management*, Longman.
Hansen, T. (1981), 'The dynamics of local expenditure growth: local government finance in Sweden', in L. J. Sharpe (ed.), *The Local Fiscal Crisis in Western Europe*, Sage.
Harlow, C. (1980), '"Public" and "Private" law: definition without distinction', *Modern Law Review*, vol. 43, no. 3.
Harris, G. M. (1939), *Municipal Self-Government in Britain*, P. S. King.
Harris, R. and Seldon, A. (1977), *Not from Benevolence*, Institute of Economic Affairs.
Harrison, M. and Norton, A. L. (1967), *Local Government Administration in England and Wales*, HMSO.
Hasluck, E. A. (1936), *Local Government in England*, Cambridge University Press.
Hatch, S. and Hadley, R. (1981), *Social Welfare and the Failure of the State*, Allen & Unwin.
Hayek, F. A. (1960), *The Constitution of Liberty*, University of Chicago Press.
Haynes, R. J. (1980), *Organisation Theory and Local Government*, Allen & Unwin.
Headey, B. (1978), *Housing Policy in the Developed Economy*, Croom Helm.
Headrick, T. (1962), *The Town Clerk in English Local Government*, Allen & Unwin.
Healey, P. (1983), 'British planning education in the 1970s and 1980s', in H. W. E. Davies and P. Healey (eds), *British Planning Practice and Planning Practice and Planning Education in the 1970s and 1980s*, Oxford Working Paper 70, Town Planning Department, Oxford Polytechnic.
Healey, P., McNamara, P., Doak, A. and Elson, M. (1985), *The Implementation of Planning Policies and the Role of Development Plans*, Report to the Department of the Environment, (unpublished) Oxford Polytechnic.
Hepworth, N. (1980), *The Finance of Local Government*, 6th edn, Allen & Unwin.
Herbert, Sir E. (Chairman) (1960), *Report of the Royal Commission on Local Government in Greater London*, Cmnd 1164, HMSO.
Hewart, Gordon (1929), *The New Despotism*, Benn.

Hicks, J. R., Hicks, U. K. and Leser, C. E. V. (1944), *The Problem of Valuation for Rating*, National Institute of Economic and Social Research, Occasional Paper 7, Cambridge University Press.
Hicks, J. R. and Hicks, U. K. (1945), *The Incidence of Local Rates in Great Britain*, National Institute of Economic and Social Research, Occasional Paper 8.
Higgins, J. *et al.* (1983), *Government and Urban Poverty*, Blackwell.
Hildersley, S. H. and Nottage, R. (1968), *Sources of Local Revenue*, Royal Institute of Public Administration.
Hill, M. J. (1972), *The Sociology of Public Administration*, Weidenfeld & Nicolson.
Hinden, R. (1950), *Local Government and the Colonies*, Allen & Unwin.
Hiscocks, R. (1957), *Democracy in Western Germany*, Oxford University Press.
Home Office (1965), *Immigration from the Commonwealth*, Cmnd 2739, HMSO.
Home Office (1975), *Racial Discrimination*, Cmnd 6234, HMSO.
Hooper, A. (1985) 'Land availability studies and private housebuilding', in S. M. Barrett and P. Healey (eds), *Land Policy: Problems and Alternatives*, Gower
House of Commons (1933a), *Debates*, 1932–3, 6 July 1933, vol. 280, col. 540.
House of Commons (1933b), *Debates*, 1932–3, 7 July 1933, vol. 280, col. 658–9.
House of Commons (1944), *Debates*, 5th series, 3 August 1944, vol. 402, col. 1579.
House of Commons (1945), *Debates*, 1945–6, 5th series, vol. 414, col. 1222.
House of Commons (1947), *Standing Committee Official Report*, 16th December 1947, col. 157.
House of Commons (1949), *Debates*, 5th series, 25 March 1949, vol. 468, col. 74.
House of Commons (1982), Environment Committee Second Report, *Enquiry into the Methods of Financing Local Government in the Context of the Government Green Paper*, Cmnd 8449, HC 217–2, HMSO.
House of Lords (1983), *Debates*, 1982–3, 26 October 1983, vol. 444, cols. 253–75, 184–357.
Housing Research Group (1981), *Could Local Authorities be Better Landlords?* Housing Research Group, City University.
Hume, D. (1984), 'Anti-poverty strategies', *Poverty*, no. 57, April.
Jackman, R. (1979), 'Monetarism and control of local spending', *CES Review*, no. 7, September.
Jackman, R. (1984), 'The Rates Bill – a measure of desperation' *Political Quarterly*, vol. 55, no. 2, April/June.
Jenkin, P. (1983), Secretary of State for the Environment, Speech to the Association of Metropolitan Authorities, 21 September.
Jenkyns, R. (1980), *The Victorians and Ancient Greece*, Blackwell.
Jennings, W. I. (1931), *Local Government in the Modern Constitution*, C. Knight.
Jennings, W. I. (1932), 'The Report on Ministers' Powers', *Public Administration*, vol. 10, no. 4.
Jennings, W. I. (1933), *The Law and the Constitution*, University of London Press.
Jennings W. I. (1935a), 'Central control', in H. J. Laski, W. I. Jennings and W. A. Robson (eds), *A Century of Municipal Progress 1835–1935*, Allen & Unwin.
Jennings, W. I. (1935b), 'In praise of Dicey 1885–1935', *Public Administration*, vol. 13, no. 2.
Jennings, W. I. (1936), 'The courts and administrative law – the experience of English housing legislation', *Harvard Law Review*, vol. 49, no. 3.
Jennings, W. I. and Griffith, J. A. G. (1960), *Principles of Local Government Law*, 4th edn, University of London Press.

Johnson, N. and Cochrane, A. (1981), *Economic Policy-Making by Local Authorities in Britain and West Germany*, Allen & Unwin.
Johnson, R. W. (1972), 'The nationalisation of English Rural Politics: Norfolk Southwest 1945–1970', *Parliamentary Affairs*, vol. 26, no. 1.
Jones, A. H. M. (1938), *The Greek City from Alexander to Justinian*, Clarendon.
Jones, G. W. (1969), *Borough Politics*, Macmillan.
Jones, G. W. (ed.) (1980), *Central–Local Relations in Britain*, Gower
Jones, G. W. and Norton, A. (eds) (1979), *Political Leadership in Local Authorities*, Institute of Local Government Studies.
Jones, G. W. and Stewart, J. D. (1982), 'The Layfield analysis applied to central–local relations under the Conservative government', *Local Government Studies*, vol. 8, no. 3.
Jones, G. W. and Stewart, J. D. (1984), *The Case for Local Government*, Allen & Unwin.
Jones, R. (1981), *Local Government Audit Law*, HMSO.
Jowell, J. (1977), 'Bargaining in development control', *Journal of Planning and Environment Law*.
Kasfir, N. (1983), 'Designs and dilemmas: an overview', in P. Mawhood (ed.), *Local Government in the Third World: The Experience of Tropical Africa*, Wiley.
Keeble, L. (1959), *Principles and Practice of Town and County Planning*. 2nd edn, Estates Gazette.
Keith-Lucas, B. and Richards, P. G. (1978), *A History of Local Government in the Twentieth Century*, Allen & Unwin.
Keynes, J. M. (1936), *The General Theory of Employment, Interest and Money*, Macmillan.
Labour Party (n.d.), *The Future of Local Government: The Labour Party's Post-War Policy*, Labour Party.
Labour Party National Executive Committee (1969), *Labour's Social Strategy*, Labour Party.
Laski, H. J. (1925), 'The pluralistic state', *Philosophical Review*, vol. 28.
Laski, H. J. (1926), 'Judicial review of social policy in England', *Harvard Law Review*, vol. 39, no. 7.
Laski, H. J. (1930), *Liberty in the Modern State*, Faber.
Laski, H. J. (1932), 'Note by Prof. Laski on the judicial interpretation of statutes', *Report of the Committee on Ministers' Powers, Annex V*, Cmd 4060.
Laski, H. J. (1943), *Reflections on the Revolution of our Time*, Allen & Unwin.
Laski, H. J., Jennings, W. I. and Robson, W. A. (1935), *A Century of Municipal Progress: 1835–1935*, Allen & Unwin.
Layfield, F. (Chairman) (1976), *Report of the Committee of Enquiry into Local Government Finance*, Cmnd 6453, HMSO.
Lee, J. M. (1963), *Social Leaders and Public Persons*, Oxford University Press.
Leiken, L. (1972), 'Governmental schemes for the metropolis and the implementation of metropolitan change', *Journal of Urban Law*, vol. 49, no. 4.
LGBC (Local Government Boundary Commission) (1947), *Report*, London.
Lipman, V. D. (1949), *Local Government Areas 1834–1945*, Blackwell.
Lipsey, D. (1982), 'Labour's new (non-manual) breed of councillor', *Sunday Times*, 19 September.
Livingstone, K. (1982), Report of speech to the Chartered Institute of Public Finance and Accountancy, *Public Finance and Accountancy*, September.
Loughlin, M. (1978), 'Procedural fairness: a study of the crisis in administrative law theory', *University of Toronto Law Journal*, vol. 28, no. 2.

Loughlin, M. (1981), 'Planning gain: law, policy and practice', 1 *Oxford Journal of Legal Studies*, vol. 1, no. 1.

Loughlin, M. (1982), 'Recent developments in central–local government fiscal relations', *Journal of Law and Society*, vol. 9, no. 2.

Loughlin, M. (1983), *Local Government, the Law and the Constitution*, Local Government Legal Society Trust.

Loughlin, M. (1985), 'Apportioning the infrastructure costs of urban land development', in S. M. Barrett and P. Healey (eds), *Land Policy: Problems and Alternatives*, Gower.

Madge, C. (1943), *Industry after the War*, Pilot Press.

Mair, D. (1983), *Keynes vs. Churchill: Fifteen All*, Department of Economics Working Paper, Heriot-Watt University.

Mallaby, Sir G. (1967), *Report of the Committee on the Staffing of Local Government*, HMSO.

Malpass, P. and Murie, A. (1982), *Housing Policy and Practice*, Macmillan.

Mandelker, D. R., Netsch, D. C. and Salsich, P. W. (1983), *State and Local Government in the Federal System*, 2nd edn, Michie.

Marks, S. (1982), 'Law and local authorities: counsels' opinions on budgets and rents', *Public Money*, vol. 2, no. 1.

Marriott, O. (1969), *The Property Boom*, Pan.

Marshall, T. H. (1970), in W. A. Robson and B. Crick (eds), *The Future of the Social Services*, Penguin.

Maud, J. P. R. (1932), *Local Government in Modern England*, Butterworth.

Maud, J. P. R. (1967), *Report of the Committee on the Management of Local Government*, HMSO.

Maud, J. P. R. and Finer, S. E. (1953), *Local Government in England and Wales*, Oxford University Press.

Mawhood, P. (ed.) (1983a), *Local Government in the Third World: The Experience of Tropical Africa*, Wiley.

Mawhood, P. (1983b), 'Decentralisation: the concept and the practice', in P. Mawhood (ed.), *Local Government in the Third World: The Experience of Tropical Africa*, Wiley.

May, Sir G. (Chairman) (1931), *Report of the Committee on National Expenditure*, Cmd 3920, HMSO.

McAuslan, J. P. W. B. (1974), 'Planning law's contribution to the problems of an urban society', *Modern Law Review*, vol. 37, no. 2.

McAuslan, J. P. W. B. (1981), *The Ideologies of Planning Law*, Pergamon.

McCallum, J. D. (1980) 'Statistical trends of the British conurbations', in G. C. Cameron (ed.), *The Future of the British Conurbations*, Longman.

McKeown, T. and Lowe, C. R. (1974), *An Introduction to Social Medicine*, 2nd edn, Blackwell.

Mény, Y. (1984), 'Decentralisation in socialist France: the politics of pragmatism', *West European Politics*, vol. 7, no. 1.

Merrett, S. (1979), *State Housing in Britain*, Routledge & Kegan Paul.

Millward, R. and Parker, D. M. (1983), 'Public and private enterprise: comparative behaviour and relative efficiency' in R. Millward, D. Parker, L. Rosenthal, M. T. Sumner and N. Topham, *Public Sector Economics*, Longman.

Milner-Holland, Sir E. (1965), *Report of the Committee on Housing in Greater London*, Cmnd 2605, HMSO.

Ministry of Health (1931), *Circular Letter*, no. 1222, 11 September.

Ministry of Health (1932), *Thirteenth Report of the Minister of Health, 1931–2*, Cmd 4113, HMSO.

Ministry of Health (1933), *Fourteenth Report of the Minister of Health, 1932–3*, Cmnd 4372, HMSO.
Ministry of Housing and Local Government (1969), *People and Planning*, HMSO.
Morrison, H. (1935), *How Greater London is Governed*, Lovat, Dickson & Thompson.
Murie, A. (1983), *Housing Inequality and Deprivation*, Heinemann.
NALGO (1933a), *Report of the Emergency Sub-Committee*, 23 January.
NALGO (1933b), *Report of the Emergency Sub-Committee*, 27 January.
NALGO (1933c), *Report of the Emergency Sub-Committee*, 8 March.
NALGO (1933d), *Proceedings of the Annual Conference, 3–5 June*, NALGO.
NALGO (1935), *NALGO During 1935*, NALGO.
NALGO (1936), *Proceedings of the Annual Conference, 30th May–1st June*, NALGO.
Nathan, R., Doolittle, F. and associates (1983), *The Consequences of Cuts: The Effects of the Reagan Domestic Program on State and Local Governments*, Urban and Regional Research Center, Princeton University, NJ.
National Consumer Council (1976), *Tenancy Agreement between Councils and their Tenants*, National Consumer Council.
Newcomer, M. (1937), *Central and Local Finance in Germany and England*, Columbia University Press.
Newton, K. (ed.) (1980), *Urban Political Economy*, Frances Pinter.
Oates, W. E. (1972), *Fiscal Federalism*, Harcourt, Brace.
O'Connor, J. R. (1973), *The Fiscal Crisis of the State*, St Martin's Press.
Onslow, R. W. A. (Chairman) (1929), *Second Report of the Royal Commission on Local Government*, Cmd 3213, HMSO.
Oyugi, N. O. (1983), 'Local government in Kenya: a case of institutional decline', in P. Mawhood (ed.), *Local Government in the Third World: The Experience of Tropical Africa*, Wiley.
Page, H. R. (1936), *Co-ordination and Planning in the Local Authority*, Manchester University Press.
Peacock, A. T. and Wiseman, J. (1967), *The Growth of Public Expenditure in the United Kingdom*, revised edn, Allen & Unwin.
Pinder, J. (ed.) (1982), *Fifty Years of PEP, 1931–1981*, Heinemann.
Plowden, Lady (1967), *Children and their Primary Schools*, Report of the Central Advisory Council for Education (England), Department of Education and Science.
Prest, A. R. (n.d.), 'Rent control and property tenetism: the experience of the U.K.' LSE mimeo.
Prest, A. R. (1983), 'Greener still and greener', *Local Government Studies*, vol. 8, no. 3.
Ray, Sir W. (Chairman) (1932), *Report of the Committee on Local Expenditure*, Cmd 4200, HMSO.
Redcliffe-Maud, Lord (Chairman) (1969), *Report of the Royal Commission on Local Government in England*, Vol. I, Cmnd 4040, HMSO.
Regan, D. E. (1977), *Local Government and Education*, Allen & Unwin.
Rhodes, G. (1970), *The Government of London: the struggle for Reform*, Weidenfeld & Nicolson.
Rhodes, R. A. W. (1981), *Control and Power in Central–Local Government Relations*, Gower.
Rhodes, R. A. W. (1984), 'Continuity and change in British central–local relations: 'the Conservative threat' 1979–83', *British Journal of Political Science*, vol. 14, no. 3.

Rhodes, R. A. W., Hardy, B. and Pudney, K. (1983a), 'The reorganisation of local government and the national community of local government', Discussion Paper 4, SSRC Central–Local Relations Project.
Rhodes, R. A. W., Hardy, B. and Pudney, K. (1983b), 'Power dependence theories of central–local relations: a critical assessment', Discussion Paper 7, SSRC Central–Local Relations Project.
Robson, W. A. (1928), *Justice and Administrative Law*, Stevens.
Robson, W. A. (1931a), *The Development of Local Government*, Allen & Unwin.
Robson, W. A. (1931b), Letter to the Editor, *Spectator*, 13 February 1931.
Robson, W. A. (1932a), 'The Report of the Committee on Ministers' Powers', *Political Quarterly*, vol. 3, p. 346.
Robson, W. A. (1932b), *The Arithmetic of the Rates*, NALGO.
Robson, W. A. (1933), 'The central domination of local government', *Political Quarterly*, January–March.
Robson, W. A. (1935a), 'The public utilities services', in H. J. Laski, W. I. Jennings and W. A. Robson (eds), *A Century of Municipal Progress*, Allen & Unwin.
Robson, W. A. (1935b), 'The outlook', in *Municipal Progress*, pp. 445–64.
Robson, W. A. (1935c), *Civilisation and the Growth of Law*, Macmillan.
Robson, W. A. (1940), *Labour Under Nazi Rule*, Oxford Pamphlets.
Robson, W. A. (1948), *The Development of Local Government*, 2nd edn (revised and enlarged), Allen & Unwin
Robson, W. A. (1950), 'Harold Laski', *Public Administration*, autumn.
Robson, W. A. (1951), *Justice and Administrative Law*, 3rd edn, Stevens.
Robson, W. A. (1959), 'Administrative law', in M. Ginsberg (ed.), *Law and Opinion in England in the Twentieth Century*, Stevens.
Robson, W. A. (1966), *Local Government in Crisis*, Allen & Unwin.
Robson, W. A. (1971), *The Political Quarterly in the 1930s*, Allen Lane.
Robson, W. A. (1979), 'Justice and administrative law reconsidered', *Current Legal Problems*, vol. 32.
Rose, E. J. B., *et al.* (1969), *Colour and Citizenship*, Oxford University Press.
Rose, R. (1980), *Politics in England*, Faber.
Royal Institute of Public Administration (1956), *New Sources of Local Revenue*, Allen & Unwin.
Samuel, G. (1983), 'Public and private law: a private lawyer's response', *Modern Law Review*, vol. 46, no. 5.
Saunders, P. (1979), *Urban Politics: A Sociological Interpretation*, Hutchinson.
Saunders, P. (1981), *Social Theory and the Urban Question*, Hutchinson.
Sayre, W. (1973), 'The relevance of the Greater London governmental experience to New York City government', in State Study Commission for New York City, *The Neighborhoods, The City, and The Region: Working Papers on Jurisdiction and Structure*.
Scarman, Lord (1981), *The Brixton Disorders, 10–12 April 1981*, Cmnd 8427, HMSO.
Schofield, M. (1977), 'The "nationalisation" of local politics', *New Society*, 28 April.
Scott, Lord (Chairman) (1942), *Report of the Committee on Land Utilisation in Rural Areas*, Cmnd 6378, HMSO.
Seebohm, Lord (Chairman) (1968), *Report of the Committee on Local Authority and Allied Personal Social Services*, Cmnd 3703, HMSO.
Sharpe, L. J. (1976), 'Instrumental participation and urban government', in J. A. G. Griffith (ed.), *From Policy to Administration: Essays in Honour of William A. Robson*, Allen & Unwin, pp. 115–38.

Sharpe, L. J. (ed.) (1981a), *The Local Fiscal Crisis in Western Europe: Myths and Realities*, Sage.
Sharpe, L. J. (ed.) (1981b) 'Is there a fiscal crisis in western European government?', in L. J. Sharpe (ed.), *The Local Fiscal Crisis in Western Europe*, Sage.
Sharpe, L. J. (1983), 'The co-ordination of sub-national government and the policy network', SSRC Conference on Political Theory and Central–Local Relations, Nuffield College, Oxford, 18 September.
Sharpe, L. J. and Newton, K. (1984), *Does Politics Matter?*, Oxford University Press.
Simon, Sir E. (1935), 'Housing and civic planning', in H. J. Laski, W. I. Jennings and W. A. Robson (eds), *A Century of Municipal Progress 1835–1935*, Allen & Unwin.
Simon, H. A. (1957), *Administrative Behaviour*, Macmillan.
Skelcher, C. K. (1982), 'Corporate planning in local government', in S. Leach and J. D. Stewart (eds), *Approaches to Public Policy*, Allen & Unwin.
Skidelsky, R. (1967), *Politicians and the Slump: the Labour Government of 1929–1931*, Macmillan.
Smellie, K. B. (1957), *A History of Local Government*, Allen & Unwin.
Smith, F. (1935), 'The nation's schools', in H. J. Laski, W. I. Jennings and W. A. Robson (eds), *A Century of Municipal Progress 1835–1935*, Allen & Unwin.
Sofer, A. (1983), 'Palmy days for council counsel', *The Times*, 15 August.
SOLACE (1976), *Employee Participation and Local Government*, Society of Local Authority Chief Executives.
Spoor, A. (1967), *White Collar Union: Sixty Years of NALGO*, Heinemann.
State Study Commission for New York City (1973), *The Neighborhoods, The City, and The Region: Working Papers on Jurisdiction and Structure*.
Steadman, F. (1932), 'Economy in purchase and in use', *Public Administration*. January.
Sternlieb, G. and Hughes, J., (1977), 'New regional and metropolitan realities of America', *Journal of the American Institute of Planners*. vol. 43, no. 3.
Stevenson, J. (1984), *British Society 1914–45*, Penguin.
Stewart, J. D. *et al.* (1974), *Local Government: Approaches to Urban Deprivation*, Home Office Urban Deprivation Unit Occasional Paper 1.
Stewart, J. D. (1983), *Local Government: the Conditions of Local Choice*, Allen & Unwin.
Stewart, M. (1980), *Life and Labour: an Autobiography*, Sidgwick & Jackson.
Sunkin, M. (1983), 'Judicial review: rights and discretion in public law', *Modern Law Review*, vol. 46, no. 5.
Task Force on Jurisdiction and Structure, State Study Commission for New York City (1972), *Re-Structuring the Government of New York City*, March.
Tavern, E. E. (1976), 'Review of *The Story of Clay Cross* by D. Skinner and J. Langdon', *Local Government Studies*, vol. 2, no. 1.
Taylor, A. J. P. (1965), *English History, 1914–1945*, Clarendon.
Temporary Commission on City Finances (1978), *The City in Transition: Prospects and Policies for New York*, Arno Press.
Teubner, G. (1983), 'Substantive and reflexive elements in modern law', *Law and Society Review*, vol. 17, pt 2.
Teubner, G. (ed.) (1984), *Dilemmas of Law in the Welfare State*, Walter de Gruyter.
Thompson, F. M. L. (1963), *English Landed Society in the Nineteenth Century*, Routledge & Kegan Paul.
Thornbury, R. (1978), *The Changing Urban School*, Methuen.

Titmuss, R. M. (1938), *Poverty and Population*, Macmillan.
Titmuss, R. M. (1950), *Problems of Social Policy*, HMSO.
Titmuss, R. M. (1963), *Essays on the Welfare State*, Unwin University Books.
Treasury (1931), *Memorandum on the Measures Proposed by His Majesty's Government to Secure Reductions in National Expenditure*, Cmnd 3952, HMSO.
Treasury (1984), *The Government's Expenditure Plans 1984/5 to 1986/7*, Cmnd 9143, HMSO.
Trubek, D. (ed.) (1983), *Reflexive Law and the Regulatory Crisis*, Madison.
US Advisory Commission on Intergovernmental Relations (1973), *City Financial Emergencies: The Intergovernmental Dimensions*, ACIR.
Urwick, L. (1935), 'Executive decentralisation with functional co-ordination', *Public Administration*, October.
Vandlandingham, K. E. (1968), 'Municipal home rule in the United States', *William and Mary Law Review*, vol. 10, no. 2.
von Beyme, K. (1983), *The Political System of the Federal Republic of Germany*, Gower.
Wade, H. W. R. (1982), *Administrative Law*, 5th edn, Oxford University Press.
Wald, K. (1977), 'The rise of class-based voting in London', *Comparative Politics*, vol. 9, no. 2.
Walker, D. (1983), 'Local interest and representation: the case of "class" interest among Labour representatives in inner London', *Government and Policy*, vol. 1.
Walsh, K. (1982), 'Local government militancy in Britain and the United States', *Local Government Studies*, vol. 8, no. 6.
Webb, Sidney (1920), *Grants in Aid*, Longmans Green.
Webb, Sidney and Webb, Beatrice (1920), *A Constitution for the Socialist Commonwealth of Great Britain*, the authors.
Weinstein, B. L. and Firestine, R. E. (1978), *Regional Growth and Decline in the United States: The Rise of the Sunbelt and the Decline of the Northeast*, Praeger.
White, M. and White, L. (1962), *The Intellectual Versus the City*, Harvard University Press.
Wiseman, H. V. (1967), *Local Government at Work*, Routledge & Kegan Paul.
Wolman, H. (1981), *The Fiscal Consequences of Urban Decline in OECD Countries*, Urban Institute.
Wolman, H. (1982), 'Local autonomy and intergovernmental finance in Britain and the United States', in R. Rose and E. Page (eds), *Fiscal Stress in Cities*, Cambridge University Press.
Wright, D. S. (1974), 'Intergovernmental relations: an analytical overview', *The Annals of the American Academy of Political and Social Science*, no. 416.
Young, K. (1975), *Local Politics and the Rise of Party*, Leicester University Press.
Young, K. (1977), '"Values" in the policy process', *Policy and Politics*, vol. 5, no. 3.
Young, K. (ed.) (1983), *National Interests and Local Government*, Heinemann.
Young, K. and Garside, P. L. (1982), *Metropolitan London: Politics and Urban Change, 1837–1981*, Edward Arnold.
Young, K. and Kramer, J. (1978), *Strategy and Conflict in Metropolitan Housing: Suburbia Versus the Greater London Council*, Heinemann.
Young, K. and Mason, C. (eds) (1983), *Urban Economic Development*, Macmillan.
Young, K. and Mills, L. (1980), *Public Policy Research: A Review of Qualitative Methods*, Social Science Research Council.
Young, K. and Mills, L. (1983), *Managing the Post-Industrial City*, Heinemann.

Table of Cases

British Cases

Anns v. *Merton London Borough Council* (1978) AC 728; (1977) 2 WLR 1024.
Associated Provincial Picture Houses Ltd v. *Wednesbury Corporation* (1948) 1 KB 223
Attorney-General v. *Fulham Corporation* (1921) 1 Ch. 440.
Bromley London Borough Council v. *Greater London Council* (1983) 1 AC 768; (1982) 2 WLR 62.
Hall & Co. Ltd v. *Shoreham-by-Sea Urban District Council* (1964) 1 WLR 240.
Kruse v. *Johnson* (1898) 2 QB 91
Lambeth London Borough Council v. *Secretary of State for Social Services* (1980) 79 LGR 61.
O'Reilly v. *Mackman* (1983) AC 120; (1983) 3 WLR 1096.
Pickwell v. *Camden London Borough Council* (1983) QB 962; (1983) 2 WLR 383.
Prescott v. *Birmingham Corporation* (1955) 1 Ch. 210.
R. v. *Greater London Council, ex parte Royal Borough of Kensington and Chelsea* (1982) *The Times*, 7 April.
R. v. *Inland Revenue Commissioners, ex parte National Federation of Self-Employed* (1982) AC 617; (1981) 2 WLR 722.
R. v. *London Transport Executive, ex parte Greater London Council* (1983) QB 484; (1983) 2 WLR 702.
R. v. *Merseyside County Council, ex parte Great Universal Stores Ltd* (1982) 80 LGR 639.
R. v. *Secretary of State for the Environment, ex parte Brent London Borough Council* (1982) QB 593; (1982) 2 WLR 693.
R. v. *Secretary of State for the Environment, ex parte Hackney London Borough Council* (1983) 1 WLR 524.
R. v. *Secretary of State for the Environment, ex parte Norwich City Council* (1982) QB 808; (1982) 2 WLR 580.
Ridge v. *Baldwin* (1964) AC 40
Roberts v. *Hopwood* (1925) AC 578.
Secretary of State for Education and Science v. *Tameside Metropolitan Borough Council* (1977) AC 1014; (1976) 3 WLR 641.
Shelley v. *London County Council* (1949) AC 56.
Wandsworth London Borough Council v. *Winder* (1984) 3 WLR 563.

US Cases

Arceneaux v. *Treen*, 671 F.2d 128 (5th Cir. 1982).
Avery v. *Midland County*, 390 US 474 (1968).
Brown v. *Thomson*, 103 S.Ct 2690 (1983).
City of New Orleans v. *Treen*, 431, So. 2d 390 (La 1983).
Cooper v. *Aaron*, 358 US 1 (1958).
Flushing National Bank v. *Municipal Assistance Corp.*, 40 NY 2d 731, 358 NE 2d 848, 390 NYS 2d 22 (1976).
Mahan v. *Howell*, 410 US 315 (1973).
Massachusetts v. *United States*, 435 US 444 (1978).

National League of Cities v. *Usery*, 426 US 833 (1976).
Reynolds v. *Sims*, 377 US 533 (1964).
Serrano v. *Priest*, 18 Cal. 3d 728, 557 P. 2d 929, 135 Cal. Rptr. 345 (1976), *cert. denied*, 432 US 907 (1977).
State v. *City of Sunrise*, 354 So. 2d 1206 (Fla 1978).

Notes on Contributors

Alan Alexander is Lecturer in Politics at the University of Reading and is the author of *Local Government in Britain Since Reorganisation* and *The Politics of Local Government in the United Kingdom*, both of which were published in 1982. He was a member of Berkshire County Council from 1973 to 1977 and a member of the policy committee of the Association of County Councils from 1975 to 1977. He served on Reading County Borough Council from 1972 to 1974 and his history of that authority – *Borough Government and Politics: Reading 1835–1985* – will appear in 1985.

Diane A. Dawson is University Lecturer and Fellow of Corpus Christi College in the University of Cambridge and was previously lecturer in economics at the University of Glasgow. Her publications are mainly in the fields of fiscal federalism and local government finance.

Nicholas Deakin read history at Oxford and then entered the home Civil Service, which he left in 1963 to join the Institute of Race Relations, where he worked on the Nuffield Survey of Race Relations in Britain. Subsequently he spent three years at the University of Sussex before joining the GLC in 1972 as Head of Social Studies, becoming Head of the Council's Central Policy Unit in 1979. In 1980 he was appointed Professor of Social Policy and Administration at Birmingham University. He has published books and articles on race relations, inner city policy and new towns.

M. David Gelfand is C. J. Morrow Research Professor of Law and Director of the Center for Legal Studies on Intergovernmental Relations at Tulane Law School, New Orleans. During 1983 he held an ESRC fellowship on the law relating to intergovernmental relations in the UK. He is the author of many articles on intergovernmental relations and his *Federal Constitutional Law and American Local Government* was published in 1984.

Malcolm Grant is Senior Lecturer in Law at Southampton University. He is the editor of the *Encyclopedia of Planning Law and Practice* and author of the *Planning Law Handbook* (1981) and *Urban Planning Law* (1982).

J. A. G. Griffith is Emeritus Professor of Public Law at the London School of Economics. He is the author of many books including *Principles of Administrative Law* (with H. Street, 5th edn, 1973), *Central Departments and Local Authorities* (1966), *Parliamentary Scrutiny of Government Bills* (1974), *The Politics of the Judiciary* (3rd edn, 1985), *Government and Law* (with T. C. Hartley, 2nd edn, 1981) and *Public Rights and Private Interests* (1982). He was a colleague of Laski, Jennings and Robson and has edited Jennings, *Principles of Local Government Law* (4th edn, 1960) and *From Policy to Administration: Essays in Honour of William A. Robson* (1976).

Patsy Healey is Associate Head of the Department of Town Planning at Oxford Polytechnic. A geographer-planner with practice experience and many years as a teacher and researcher into planning, her recent work includes a major project for the Department of the Environment on the implementation of development plans. Her recent publications include *Local Plans in British Land Use Planning* (1983), *Land Policy: Problems and Alternatives* (edited with S. M. Barrett, 1985), and *The Political Economy of Land* (with Alan Gilbert, 1985).

John Gyford is Lecturer in Urban Politics and the Politics of Planning at University College London. He is the author of *Local Politics in Britain* (1984) and (with M. James, 1983) *National Parties and Local Politics*. He has served as a councillor on Essex County Council from 1970 to 1977 and on Braintree District Council since 1983.

Richard Jackman is Senior Lecturer in Economics at the London School of Economics. He is joint author (with C. D. Foster and M. Perlman) of *Local Government Finance in a Unitary State* (1980) and he has published articles on local government finance in *Public Choice, Local Government Studies* and *CES Review*. He is also co-author of a book on inflation and has published various papers on macroeconomic topics.

Martin Loughlin is Lecturer in Law at the London School of Economics. He has previously taught at Osgoode Hall Law School, York University, Toronto, and the University of Warwick. In 1982/3 he was a Visiting Fellow at the School for Advanced Urban Studies, University of Bristol.

Alan Murie is Lecturer in Urban Studies at the School for Advanced Urban Studies, University of Bristol. He has been involved in a wide range of research on housing and housing policy and was specialist adviser to the House of Commons Environment Committee between 1980 and 1982. His publications include *Housing Policy and the Housing System* (with P. Niner and C. Watson), *Housing Policy and Practice* (with P. Malpass, 1983) and *Housing Inequality and Deprivation* (1983).

J. D. Stewart has been Professor of Local Government and Administration at the University of Birmingham since 1971 and from 1976 to 1983 was Director of the Institute of Local Government Studies. He served on the Layfield Committee on Local Government Finance and his books include *British Pressure Groups* (1958), *Management in Local Government: a Viewpoint* (1971), *The Responsive Local Authority* (1974), *Local Government: the Conditions of Local Choice* (1983). A series of articles written by himself and Professor G. W. Jones were brought together in *The Case for Local Government*, published in 1983.

Ken Young is Senior Fellow at the Policy Studies Institute and was previously at the School for Advanced Urban Studies, University of Bristol. During 1982/3 he was specialist adviser to the House of Commons Environment Committee. His books include *Local Politics and the Rise of Party* (1975), *Metropolitan London: Politics and Urban Change, 1837–1981* (with P. L. Garside, 1982) and *Managing the Post-Industrial City* (with Liz Mills, 1983).

INDEX